On-Chip Training NPU - Algorithm, Architecture and SoC Design

AF781116

Donghyeon Han • Hoi-Jun Yoo

On-Chip Training NPU - Algorithm, Architecture and SoC Design

Springer

Donghyeon Han
Massachusetts Institute of Technology
Cambridge, MA, USA

Hoi-Jun Yoo
Department of Electrical Engineering
KAIST
Daejeon, Korea (Republic of)

ISBN 978-3-031-34239-4 ISBN 978-3-031-34237-0 (eBook)
https://doi.org/10.1007/978-3-031-34237-0

© The Editor(s) (if applicable) and The Author(s), under exclusive license to Springer Nature Switzerland AG 2023
This work is subject to copyright. All rights are solely and exclusively licensed by the Publisher, whether the whole or part of the material is concerned, specifically the rights of translation, reprinting, reuse of illustrations, recitation, broadcasting, reproduction on microfilms or in any other physical way, and transmission or information storage and retrieval, electronic adaptation, computer software, or by similar or dissimilar methodology now known or hereafter developed.
The use of general descriptive names, registered names, trademarks, service marks, etc. in this publication does not imply, even in the absence of a specific statement, that such names are exempt from the relevant protective laws and regulations and therefore free for general use.
The publisher, the authors, and the editors are safe to assume that the advice and information in this book are believed to be true and accurate at the date of publication. Neither the publisher nor the authors or the editors give a warranty, expressed or implied, with respect to the material contained herein or for any errors or omissions that may have been made. The publisher remains neutral with regard to jurisdictional claims in published maps and institutional affiliations.

This Springer imprint is published by the registered company Springer Nature Switzerland AG
The registered company address is: Gewerbestrasse 11, 6330 Cham, Switzerland

Abstract

Deep learning becomes the mainstream of artificial intelligence applications and its demand is increasing day by day. Previously, deep learning was only considered for cloud-server applications because of the huge computation amount. Many edge/mobile devices are now able to utilize deep neural networks (DNNs) thanks to the development of mobile DNN accelerators. Mobile DNN accelerators overcame the problem of limited computing resources and battery capacity by realizing energy-efficient inference. However, it shows passive behavior so it makes AI hard to be actively interacting with individual users or its service environment. The importance of on-chip training becomes rising more and more because of this limitation. Despite its advantages, DNN training has more constraints than inference, so it was hard to be realized in mobile/edge devices. This book suggests typical DNN training processors: DF-LNPU, which is focused on the acceleration of a specific application; and HNPU, a general-purpose DNN training processor, which is an energy-efficient DNN training processor by adopting algorithm-hardware co-design. The first DNN training processor, DF-LNPU, is proposed using direct feedback alignment (DFA). The proposed processor achieves a 2.2 $\times$ faster DNN training speed compared with the previous processors by the pipelined DFA. In order to enhance the energy efficiency by 38.7%, the heterogeneous learning core architecture is optimized with the 11-stage pipeline data path. Furthermore, the direct error propagation core utilizes random number generators to remove external memory access caused by error propagation and improves the energy efficiency by 19.9%. The DF-LNPU is evaluated on the object tracking application, and as a result, it shows 34.4 frames-per-second throughput with 1.32 TOPS/W energy efficiency. The second processor, HNPU, realizes energy-efficient DNN training by focusing on general-purpose DNN training. It supports stochastic dynamic fixed-point representation and layer-wise adaptive precision searching unit for low-bit-precision training. It additionally utilizes slice-level reconfigurability and sparsity to maximize its efficiency both in DNN inference and training. Adaptive-bandwidth reconfigurable accumulation network enables reconfigurable DNN allocation and maintains its high core utilization even in various bit-precision conditions. Fabricated in a 28-nm process, the HNPU accomplished at least 5.9 $\times$

higher energy efficiency and 2.5 $\times$ higher area efficiency in general DNN training benchmark such as ImageNet compared with the previous state-of-the-art on-chip learning processors.

Keywords

Deep learning, Deep neural network training, Digital processor, Energy-efficient hardware, Back-propagation, Backward locking, Direct feedback alignment, On-chip learning, Sparsity exploitation, Adaptive precision searching

Contents

List of Figures

List of Tables

Chapter 1
Introduction

Artificial intelligence (AI) is already essential in the current smart machines. For example, smartphones now utilize AI-based face recognition for user authentication. Their AI voice assistant can improve the user's convenience because it can remove the user's manual device operation. Still, AI applications used in commercial products highly depend on the cloud server. However, communication with the cloud can cause not only the delay due to unreliable latency of the communication channel but also a privacy problem because it transfers user data regardless of the user's intent (Figs. 1.1, 1.2, and 1.3).

As shown in Fig. 1.4, many deep neural network (DNN) accelerators [1–13] provided on-device inference to solve this problem. However, they give limited applications with the lower accuracy because it just repeats pre-determined task based on pre-trained DNN. The user environment cannot be easily trained during the pre-training, and the inference processor shows poor performance if it is placed in an unexpected situation. Finally, the passive behavior of AI processors prevents providing user-friendly AI services.

To make AI more interactive, demand for on-chip training has been increased. On-chip training can not only compensate for accuracy degradation but also give personalization functionality by tuning global knowledge to be optimized to the user-specific datasets. In addition, it can divide the workload of the server through the distributed learning. The distributed learning can protect even the privacy during the dataset collection so that it can easily scale up the DNN training.

Even though the on-chip DNN training can have many advantages, the realization of training at the micro-AI systems seems challenging. Specifically, edge/mobile devices must handle training with their limited computing resources and battery capacity. Since slow training cannot actively adapt itself to the ever-changing environment, training should be performed reasonably quickly and efficiently.

© The Author(s), under exclusive license to Springer Nature Switzerland AG 2023

D. Han, H.-J. Yoo, *On-Chip Training NPU - Algorithm, Architecture and SoC Design*,
https://doi.org/10.1007/978-3-031-34237-0_1

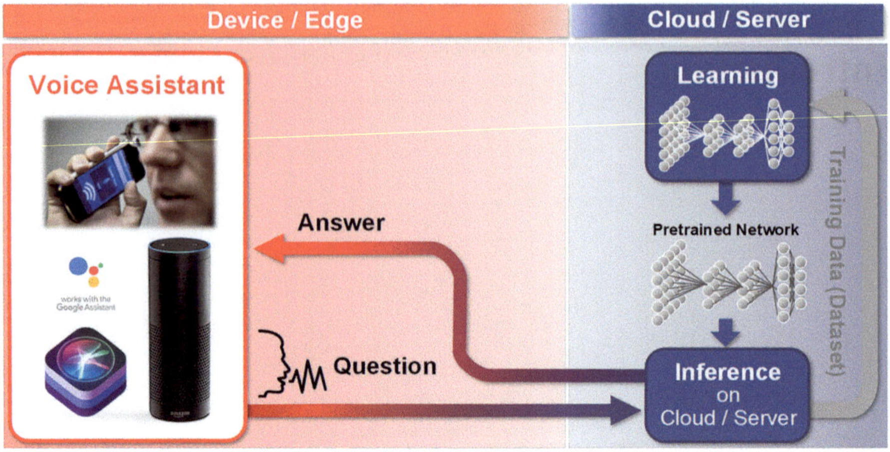

Fig. 1.1 Cloud-centric deep learning application

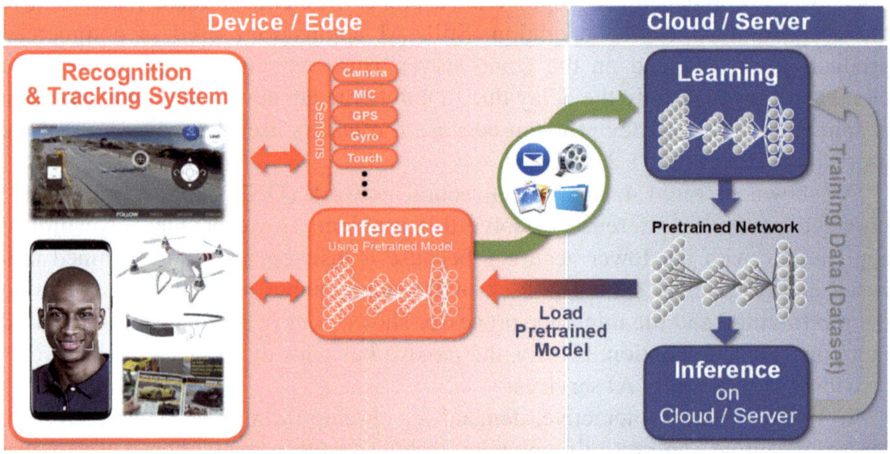

Fig. 1.2 Moving DNN inference applications to edge/mobile devices

1.1 Necessity of On-Device Deep Neural Network Training

The work [161] summarizes the six-level rating for AI which illustrates the direction of AI processors with the advancement of their intelligence. As mentioned before, initial DNN application highly depends on cloud computing, and thus, both training and inference of DNN are performed in the cloud. The level 1 intelligence loosens cloud-dependent DNN acceleration by adopting cloud–edge co-acceleration of DNN inference. The intelligence can be advanced further by making the system to have more independence of inference. These level 1–3 intelligence still leave the training part to cloud service and support only DNN inference based prediction

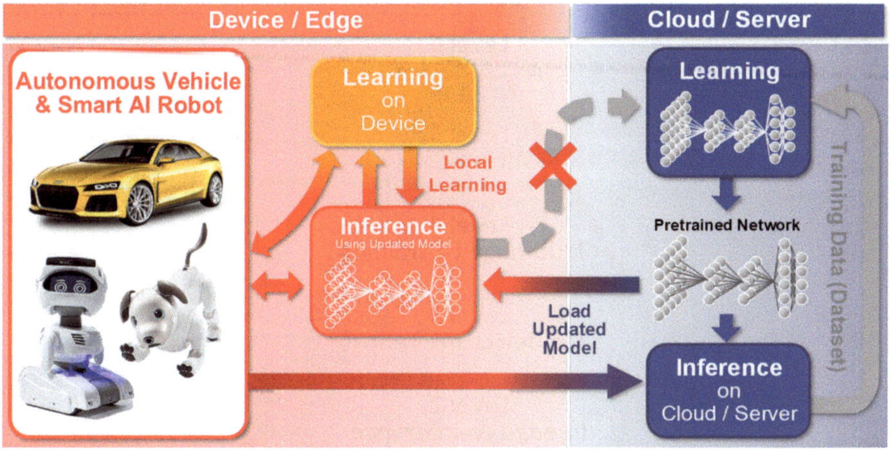

Fig. 1.3 Moving both DNN inference and training application to edge/mobile devices

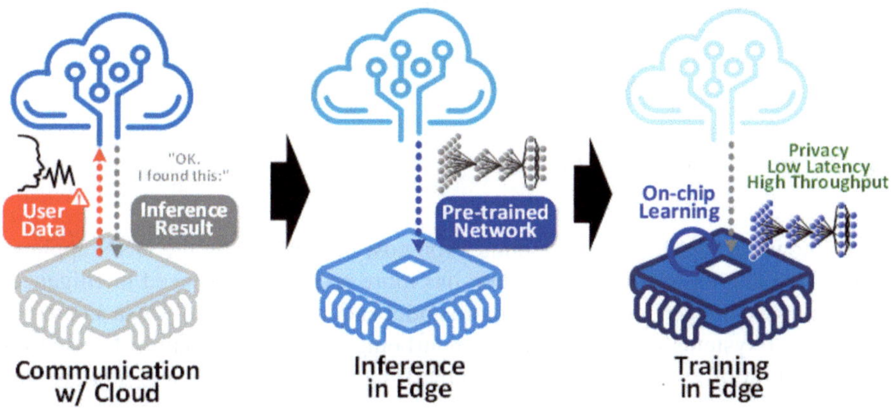

Fig. 1.4 History of DNN accelerators

applications on mobile and edge devices. The upcoming evolution of AI is DNN training on mobile or in edge devices. Similar to level 1, the easiest way to support training on mobile is cloud–edge co-training paradigm. Edge devices mainly collect data from user- or device-specific environment and cloud server transfers new DNN parameters after training. There is another type of co-training called federated learning [121]. In federated learning, each edge device trains its own models by using a local dataset and each of the trained parameters is transferred to cloud to aggregate the intelligence and makes bigger DNN which has high generality. In level 5 intelligence, both inference and training are eventually accelerated with the help of cloud service. The difference between level 5 and level 6 is portability and energy constraint because on-device training should be accelerated within mobile devices which have limited battery capacity.

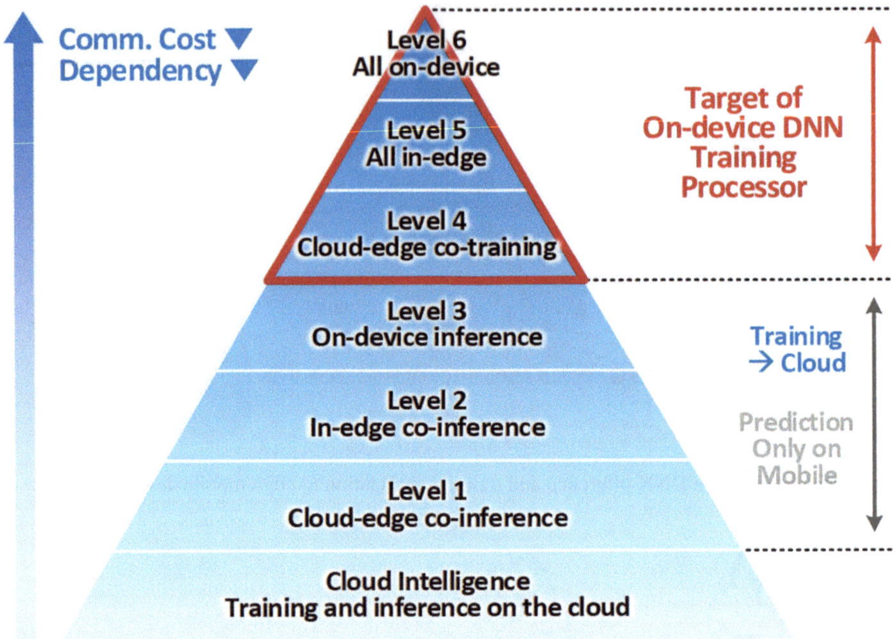

Fig. 1.5 Six-level rating for AI [161]

As shown in Fig. 1.5, the communication cost needed during the DNN application will be decreased when the system has more and more advanced intelligence. It encourages the devices to have more independence and finally can construct complex AI systems or society like human beings. This work mainly focused on the realization of levels 4–6 intelligence by designing energy-efficient and fast on-device DNN training processor.

1.2 Major Applications of On-Device DNN Training

The training efficiency and speed is the key enabler of on-device training, but the detailed requirements can be varied according to the target application. We summarize three major applications of on-device DNN training as follows: (1) evolution (long-term DNN training), (2) advancement (medium-term DNN training), and (3) adaptation (short-term DNN training).

1.2.1 Evolution: Long-Term DNN Training

In the initial research of DNN, evolution is the most common example of DNN training. If we can collect the data with its proper label, a new DNN application can be created. However, there are too much data in the real world and their individual labeling seems impossible because the labeling should be modified in different applications even though it shares the same data. At the beginning of DNN application, it highly depends on big-company-centered cloud service, and thus, the labeling becomes the main barrier to application enlargement. In addition to labeling, it faces privacy problems when gathering the training data. The users do not want to share their individual data from messages to pictures. On-device DNN training can resolve this problem through long-term DNN training such as federated learning [121]. After user- and device-wise DNN training, trained weight is transferred to the cloud and it generates global DNN. This training process targets long-term training because it needs to collect enough user data and communication for parameter sharing is done infrequently. Therefore, it needs the reconfigurable logic to support training functionality but does not need to finish them quickly (Fig. 1.6).

1.2.2 Advancement: Medium-Term DNN Training

On-device training is important when the application needs user-specific solutions. Figure 1.7 shows some examples of DNN training to support user-specific tasks. In general, we can train big teacher networks by using a standard dataset such as ImageNet [81]. It contains lots of object classes, but the majority of classes are not in practical use. On-device training can make a small student network which shrinks network size and contains only user-specific classes. It can be further optimized by using quantization or weight pruning. Quantization or weight pruning is the efficient network lightening method, and it can also help fast and efficient DNN training in the edge devices. In addition to network shrinking, it can add new classes which were not considered in big teacher network training. As shown in Fig. 1.7, the

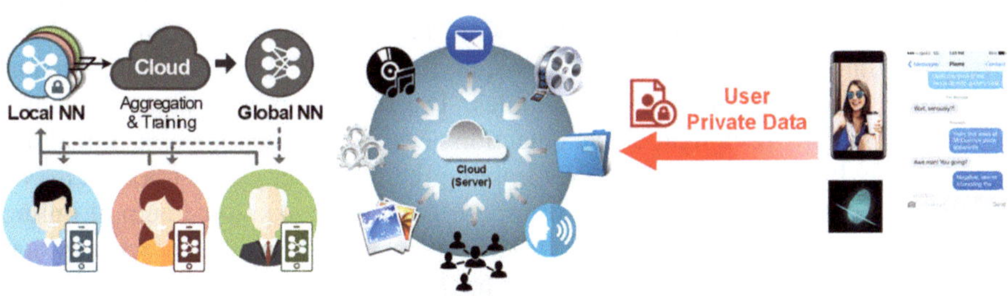

Fig. 1.6 DNN training Example 1: evolution

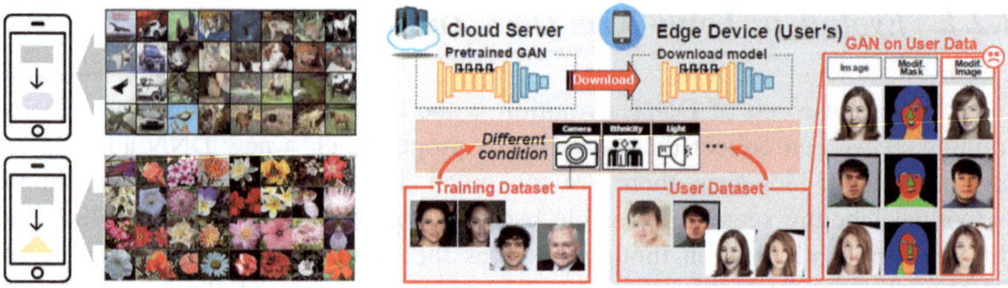

Fig. 1.7 DNN training Example 2: advancement

network trained in ImageNet can be re-trained to support flower classification. It needs to fine-tune the network which utilizes the original big teacher network as a backbone but modifies its weight in the user-specific dataset [146]. It can also solve the dataset biasing problem. Dataset biasing problem is an unavoidable problem, and thus, the performance of the DNN can be significantly degraded if the specific situation or environment cannot be covered by the pre-trained network. The authors in [32] verified that the pre-trained style-transfer generative adversarial network (GAN) was biased toward western people and shows awkward drawing when it is applied to Asian people. The work [32] also adopts on-device training-based training to recover its original performance. These applications show different on-device training requirement because they have a smaller amount of data but need to be finished during midnight which is generally considered as the device charging and not in use. The training time constraint is more restrictive than long-term DNN training cases, but it still does not need a real-time or ultra-fast performance like inference application. Instead, these training scenarios are important to offer user-friendly applications. Moreover, it can optimize networks using quantization or pruning and enables fast and efficient inference on edge/mobile devices.

1.2.3 Adaptation: Short-Term DNN Training

Both long-term and medium-term trainings are important to preserve users' privacy while expanding DNN applications to a broad area. However, their timing constraints are not that strict compared with the DNN inference scenario. However, real-time or ultra-fast training is required when DNN needs environmental adaptation. There are two typical examples. The first example is object tracking. The work [41] utilizes a light object detection network but supports object tracking functionality by adopting online DNN training. Unlike other object tracing DNNs, [41] shows a smaller size of DNN to support tracking thanks to the help of online DNN training. The corresponding network consists of a classification network that classifies whether the bounding box is our tracking target or not. Since the shape of the target object can be deformed due to its individual movement, it learns a new

Fig. 1.8 DNN training Example 3: adaptation

shape through online DNN training. Furthermore, it adapts to new illuminations or occlusions to be robust to environmental changes. Generally, object tracking targets real-time operation, and the online training process should be accelerated within 30 ms not to degrade the performance of the conventional inference process.

The other important application is adaptation after unexpected situations. The intelligence of the DNNs continues to grow as the size of the network increases but edge/mobile devices have limited memory capacity and it makes it difficult to train bigger DNNs. For this reason, edge devices inevitably have to use mobile-oriented DNNs which demand fewer parameters and computational amounts. This mobile-oriented DNN shows low generality, and thus, it could experience a fatal error in unexpected situations such as camera malfunction or abrupt domain change. Online DNN tuning performed right after an unpredictable accident shows great performance recovery [80, 144] while maintaining the hardware benefits of the lightweight DNN. Although on-device DNN training seems promising, it must be accompanied by an energy-efficient and low-latency DNN training processor. Long latency during the training causes a critical accident due to degraded performance.

Both examples require short-term training which requires the fastest training speed compared with the other two DNN training scenarios. Instead, it can loosen the generality because it does not target general-purpose applications and collects only a small amount of data which is biased to the short-term period right after an accident. In summary, short-term DNN training focuses on low-power and low-latency DNN training for real-time and ultra-fast adaptation but utilizes overfitting to follow deformable objects and overcome emergent situations (Fig. 1.8).

1.2.4 Comparison of Three Application Scenarios

We summarized the key applications that are allowed on mobile/edge devices. In the evolution scenario, it adopts long-term DNN training which utilizes big-size datasets and performs infrequent training for big teacher network generation while preserving users' individual privacy. The DNN application of the mobile-edge device can be more specified when it adopts an advancement scenario that

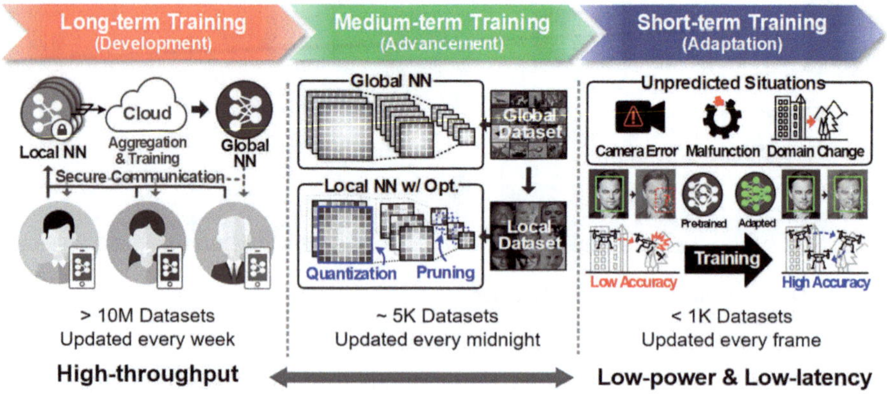

Fig. 1.9 Requirements according to DNN training types

requires medium-term DNN training to optimize DNN to a specific user or device by adopting fine-tuning and hardware-optimized training such as quantization and weight pruning. Unlike both evolution and advancement scenarios, the adaptation scenario requires ultra-fast DNN training, but it loses the generality of the original network by using only a small amount of data collected every frame (Fig. 1.9).

The design direction of the on-chip DNN training processor can be varied according to the target DNN training scenarios. In this work, we will introduce various DNN training processors with various target applications.

1.3 Contribution of This Book

This document focused on both software and hardware acceleration of DNN training on mobile devices. Before the discussion of DNN training solutions, key applications and challenges of mobile DNN training are explained in Chap. 1. Based on information summarized in Chap. 1, my key distributions in mobile DNN training field will be introduced in the following chapters. Firstly, new algorithms, binary direct feedback alignment (BDFA) and hybrid direct feedback alignment (HDFA), will be explained through Chaps. 3 and 4. Both proposed algorithms focus on resolving the backward locking problem which is a major obstacle of low-latency and energy-efficient DNN training on mobile devices. They substitute conventional back-propagation (BP) based DNN training with the direct feedback alignment (DFA) based backward unlocking method. Thanks to the direct error propagation, it can avoid sequential computation necessary during the BP. In Chap. 3, hardware-friendly DFA algorithm, BDFA, will be introduced, but it utilizes DFA-based direct error propagation only for the last few layers of DNN. In Chap. 4, the HDFA is proposed to support full-DFA based error propagation and extend its application to both convolutional neural network (CNN) and recurrent neural network (RNN).

Chapter 5 will explain a new online DNN training processor, DF-LNPU. It focused on low-latency DNN training using the new DFA algorithm illustrated in Chaps. 3 and 4. Although the DF-LNPU successfully demonstrate low-latency and energy-efficient DNN training on object tracking application, it shows low reconfigurability and generality for DNN training, and thus, it cannot be utilized for general-purpose distributed learning. In Chap. 6, general-purpose DNN training processor will be explained with the new fixed-point representation based training and bit-scalable architecture. In Chap. 7, I integrate the know-hows obtained from two algorithms and two processors to design general-purpose mobile DNN training processor but enable to support various kinds of backward unlocking solutions. The final version of ASIC named HNPU-V2 successfully demonstrated low-latency object detection only using the fixed-point representation and improves its programmability to adopt various backward unlocking solutions.

In the last chapter, Chap. 8, I will summarize and compare both conventional and proposed DNN training processors by reviewing the challenges introduced in Chap. 1. Based on summary and analysis about DNN training processor, I finally illustrate the key design differences between DNN inference and training processors on mobile devices. To sum up, there are two algorithms and three hardware contributions, and it finally suggests the major design points of mobile DNN training processor compared with the conventional inference engines.

The contents of the dissertation are composed of papers of which the first author is Donghyeon Han. The main contents are based on paper "Energy-efficient DNN Training Processors on Micro-AI Systems" (IEEE Open Journal of the Solid-State Circuits Society, [162]). Chapter 3 is written based on paper "Efficient Convolutional Neural Network Training with Direct Feedback Alignment" (ICCVW or arXiv:1901.01986 [72]), and Chap. 4 is based on paper "Extension of Direct Feedback Alignment to Convolutional and Recurrent Neural Network for Bio-plausible Deep Learning" (arXiv:2006.12830 [74]). Chapter 5 is based on paper "DF-LNPU: A Pipelined Direct Feedback Alignment-Based Deep Neural Network Learning Processor for Fast Online Learning" (IEEE Journal of Solid-State Circuits, [34]). Chapter 6 is based on papers of "HNPU: An Adaptive DNN Training Processor Utilizing Stochastic Dynamic Fixed-Point and Active Bit-Precision Searching" (IEEE Journal of Solid-State Circuits, [29]) and "A Mobile DNN Training Processor with Automatic Bit-precision Search and Fine-grained Sparsity Exploitation" (IEEE Micro [145]). Chapter 7 is based on paper "A 0.95 mJ/frame DNN Training Processor for Robust Object Detection with Real-World Environmental Adaptation" (IEEE International Conference on Artificial Intelligence Circuits and Systems [80]).

Chapter 2
A Theoretical Study on Artificial Intelligence Training

In this chapter, we will focus on the theoretical study of artificial intelligence (AI), especially for "Learning." Since "Learning" is the key enabler of the self-evolution of AI, without consideration of "Learning," AI is not that different from the conventional algorithm. Specifically, the "adaptation" part that was introduced in the short-term training scenario in Sect. 1.2 is the most important and challenging application of on-device DNN training. Before we move on to detailed solutions of "Learning" and "adaptation," I will summarize the basic principles of "Learning" in this chapter.

First, I will summarize the various types of machine learning according to labeling method, data collecting method, or the purpose of the training. It will be a great example to determine the AI application in addition to three key application scenarios introduced in Sect. 1.2. Second, the most popular and (currently known as) best training method, back-propagation, is illustrated with mathematical equations. Not only mathematical equations but also its intuitive computation format will be explained together to introduce the main challenges that on-device training processor faces. It will cover both neural networks with dense connections (multi-layer perceptron, recurrent neural network, transformer) and neural networks with sparse connections (convolutional neural network).

After an explanation of back-propagation, we will explain why back-propagation is not considered a final solution in machine learning research. The main reason is the low biological plausibility of back-propagation, which rises the importance of non-back-propagation-based training methodologies. For this reason, the other learning methods that tried to avoid back-propagation will be introduced. Although the back-propagation updates entire parameters of the neural network, partial weight update, called local learning, sometimes shows better accuracy than global learning, and it is promising research for energy-efficient and super-fast on-device training in the future. Not only gradient-free non-back-propagation algorithms but also some trials that partially utilize the back-propagation concept will also be explained. We

© The Author(s), under exclusive license to Springer Nature Switzerland AG 2023

D. Han, H.-J. Yoo, *On-Chip Training NPU - Algorithm, Architecture and SoC Design*,
https://doi.org/10.1007/978-3-031-34237-0_2

define it as semi-back-propagation, which still has a backward propagation stage for weight update but loosens the dependency to resolve the backward locking problem.

Theoretical study of AI training will be helpful to get an intuitive solution to make own on-device DNN training methodologies. Moreover, this chapter will be the theoretical background to understand the remaining chapter's idea.

2.1 Types of Machine Learning

Today's world generates huge amounts of data. In addition to being generated by humans, these data are also generated by computers, mobile phones, and other technologies. Programmers will undoubtedly decide how to train an algorithm using a particular learning model, depending on the type of data available and the motivation at hand. In the field of computer science known as machine learning, the performance of a system is improved by iteratively performing tasks using data rather than explicitly coded programs. Let us take a closer look at the differences between the three machine learning methodologies: supervised learning, unsupervised learning, and reinforcement learning (Fig. 2.1).

2.1.1 Supervised Learning

A machine learning paradigm known as supervised learning is used to solve problems when the current data consist of labeled instances, which means that each data point has a feature and a corresponding label. Supervised learning algorithms based on sample input–output pairs aim to train functions that transform function inputs into outputs. Infer a function using labeled training data that consist of a collection of training samples. Each example of supervised learning is a pair containing an input and an intended output. Functions inferred by a supervised

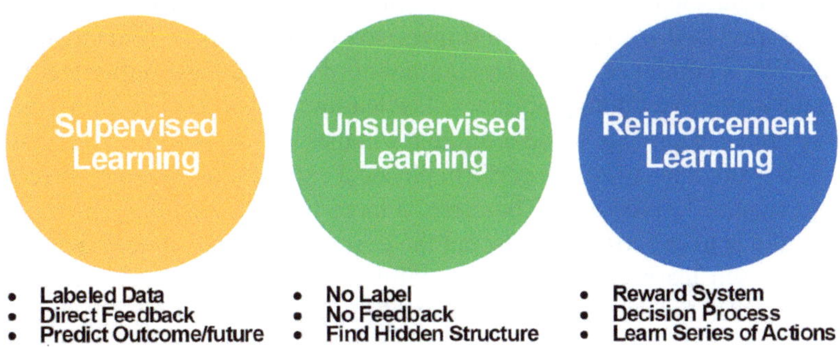

Fig. 2.1 Three main types of machine learning according to their label types: (1) supervised learning, (2) unsupervised learning, and (3) reinforcement learning

Fig. 2.2 Overview of supervised learning

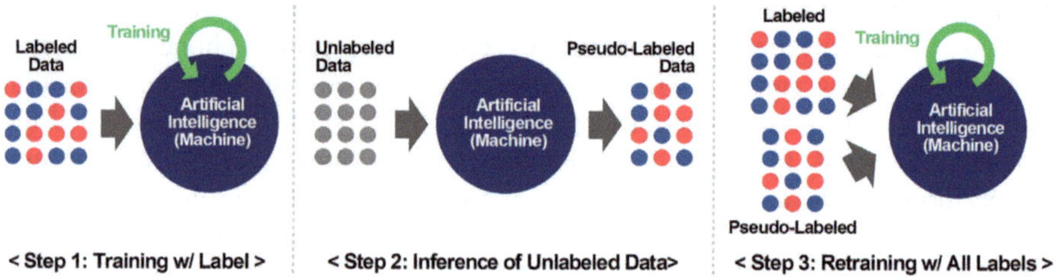

Fig. 2.3 Overview of semi-supervised learning

learning algorithm are generated from the training data, which can be used to map new samples. Algorithms can accurately determine class labels for instances that are not yet visible in an ideal environment. This requires the learning algorithm to generalize in a "reasonable" way from the training data to a hypothetical situation. The so-called generalization error is used to measure the statistical performance of an algorithm (Fig. 2.2).

Semi-supervised Learning

Semi-supervised learning is a machine learning method that mixes a significant amount of unlabeled data with a small amount of labeled data during training. Between supervised learning and unsupervised learning, there is semi-supervised learning. When combined with a small amount of labeled data, unlabeled data can significantly improve training accuracy. For learning tasks such as determining the 3D structure of a protein or determining whether an oil is present at a specific location, it is often necessary to perform physical experiments or use trained human agents to acquire labeled data. Therefore, a large fully labeled training set may be impractical due to the cost of labels, but the cost of acquiring unlabeled data is low. Semi-supervised learning can be very useful in these situations. Semi-supervised learning has the theoretical appeal of machine learning and how humans learn (Fig. 2.3).

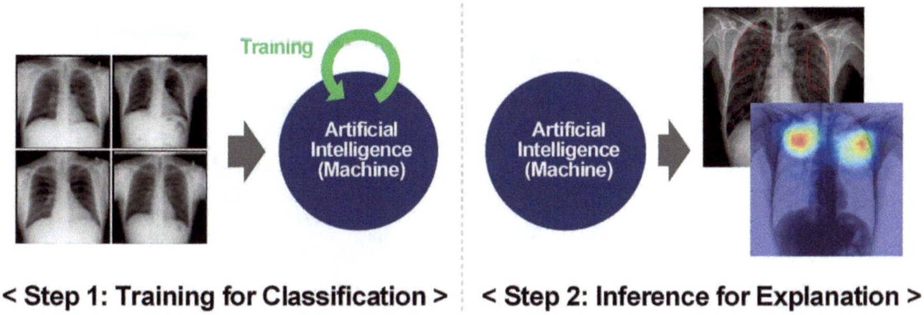

Fig. 2.4 Overview of weakly supervised learning

Weakly Supervised Learning

In the field of machine learning known as weak supervision, it provides supervisory signals for classifying huge amounts of training data using noisy, limited, or different label sources. This method reduces the difficulty of acquiring hand-labeled datasets, which can be costly or impossible to work with. Instead, cheap weak labels are leveraged with the knowledge that they can be used to build flawed but still robust predictive models (Fig. 2.4).

Self-supervised Learning

Self-supervised learning is a machine learning method that obtains knowledge from unlabeled test data. You can think of it as a kind of learning somewhere between supervised and unsupervised learning. Training a neural network involves two steps: (1) Pseudo-labels are used to initially solve the task and set the network weights. (2) Second, supervised or unsupervised learning is used to perform real-world tasks. In recent years, self-supervised learning has shown promising results, has been put to practical use in audio processing, and is now used in Facebook and other speech recognition applications. A key advantage of self-supervised learning is that it can obtain high-accuracy training results even with low-quality data. It trains on data with automatically generated labels. It itself experiences an entirely supervised learning process. However, it does not require a manually labeled dataset (Fig. 2.5).

Imitation Learning

The concept of imitation is self-explanatory. In simple terms, repeating an activity after observation. It requires a "real-time" DNN training idea that is currently difficult for AI to replicate. Or, it observes the target subjects through the sensors and memorizes them for long-term memory. This observation can be used for later

Fig. 2.5 Overview of self-supervised learning

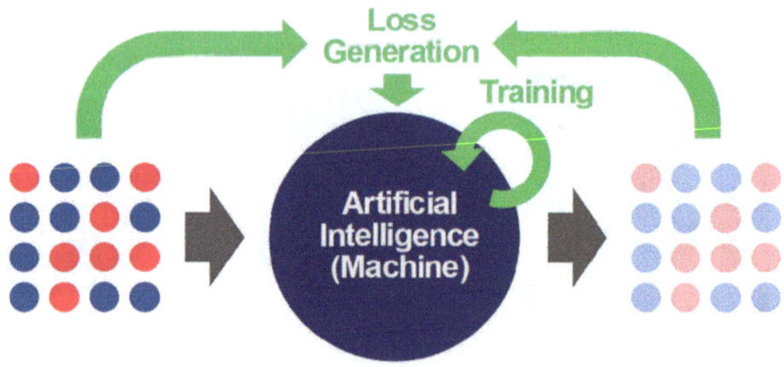

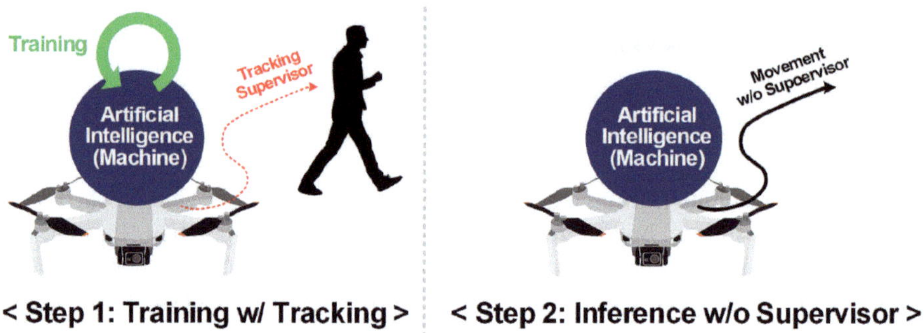

Fig. 2.6 Overview of imitation learning

DNN training without real-time training. In general, imitation learning is helpful when an expert has to define a reward function that produces the same behavior, or when the desired behavior can be more easily demonstrated than when a policy has to be explicitly learned (Fig. 2.6).

2.1.2 Unsupervised Learning

Unsupervised learning is a machine learning paradigm for problems in which a dataset consists of unlabeled samples. That is, each data point contains only features that do not have a corresponding label. Algorithms for unsupervised learning aim to discover relevant patterns or structural features in data. Dimensional reduction, density estimation, and clustering are some examples of unsupervised learning problems. Unsupervised learning aims to train machine learning algorithms to generate concise internal representations of inputs through imitation, an important human learning method. These representations can help, and perhaps be important, for an algorithm to generate creative material for generative tasks.

Moreover, current research is further developed to combine supervised learning and unsupervised learning together to enhance training performance. Unsupervised learning can offer intuitive knowledge to DNN, and it is sometimes helpful to avoid overfitting or inference in unexpected situations (Fig. 2.7).

Fig. 2.7 Overview of unsupervised learning

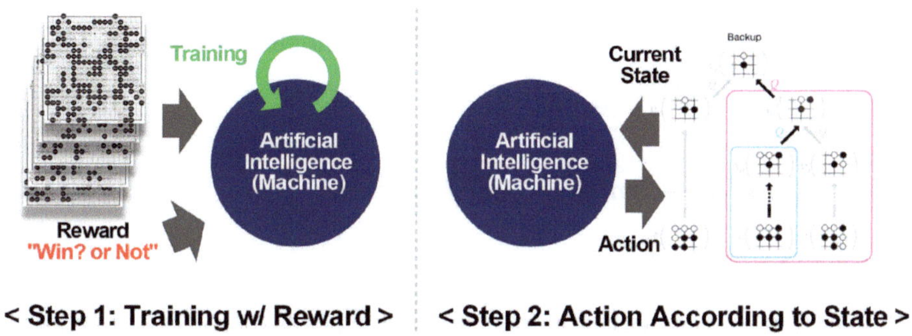

Fig. 2.8 Overview of reinforcement learning

2.1.3 Reinforcement Learning

The field of machine learning known as reinforcement learning (RL) studies how intelligent agents should behave in a given environment in order to maximize the concept of cumulative reward. Along with supervised and unsupervised learning, reinforcement learning is one of the three basic machine learning paradigms. Reinforcement learning differs from supervised learning in that it does not require explicit modification of the representation or sub-optimal behavior of labeled input–output pairs. Instead, it focuses on balancing exploitation and exploration.

Many reinforcement learning methods for these situations involve a dynamic programming approach, so the environment is usually expressed as a Markov decision process (MDP). The main difference between traditional dynamic programming techniques and reinforcement learning algorithms is that the latter does not require an understanding of the exact mathematical representation of the MDP and is more suitable for large MDPs than exact techniques. There are also partially supervised RL algorithms that effectively combine the advantages of supervised and RL methods. RL is supervised in part in that it uses a reward function designed by human experts to perform indirect supervision (Fig. 2.8).

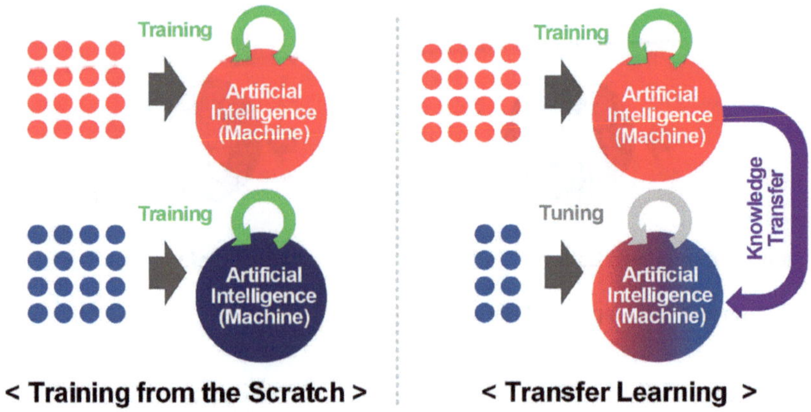

Fig. 2.9 Overview of transfer learning

2.1.4 Knowledge-Transfer-Based Learning

Transfer Learning

Transfer learning is a machine learning research topic that involves storing information obtained while solving one problem and applying it to another similar but unrelated problem. For example, skills acquired when learning to identify vehicles can be used when attempting to identify trucks. Despite the fact that there is little real connection between the two professions, this area of research has some similarities with the extensive psychological literature on learning transference. Practically speaking, the sampling efficiency of reinforcement learning agents can be significantly increased by reusing or transferring knowledge of previously learned tasks for learning new tasks. Learning a new task relies on the knowledge obtained from previously learned data; thus, it needs less training data in the current target task (Fig. 2.9).

Domain Adaptation

Domain adaptation is a subfield of machine learning and transfer learning. This situation arises when you try to apply a successful model learned from the source data distribution to a different but similar target data distribution. For example, if the network is pre-trained with real-world pictures, its inference performance can be degraded with animated or drawn subjects even with the same object. Domain adaptation recovers accuracy in this situation by adapting the model of source distribution to the new target distribution. Adaptation is an essential concept of AI because training the DNN with entirely known subjects is impossible. Adaptation is helpful to study unknown subjects based on knowledge obtained from known

Fig. 2.10 Overview of domain adaptation

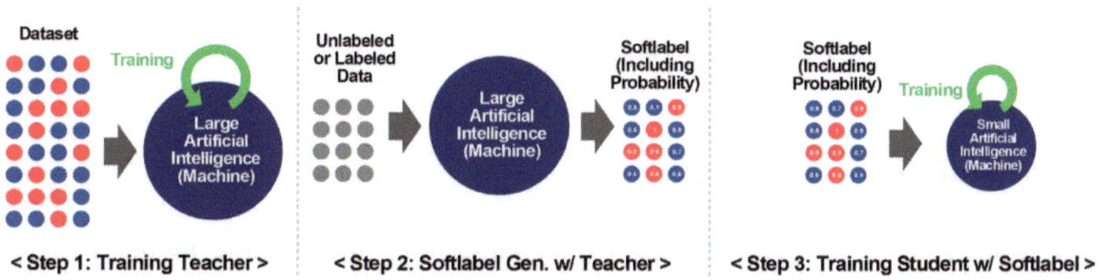

Fig. 2.11 Overview of knowledge distillation

subjects. When more than one source distribution is accessible, this issue is called multi-source domain adaptation. (Input distribution changes, but the labels remain the same) (Fig. 2.10).

Knowledge Distillation

The transfer of knowledge from a large model to a small model is called knowledge distillation in machine learning. Large models (such as very deep neural networks or ensembles of multiple models) have a greater knowledge capacity than small models, but the relative size of the teacher network is not always bigger than the student network. Without loss of validity, knowledge distillation moves information from a large model to a small model. Knowledge distillation usually utilizes soft label that indicates not only label type but also its detailed inference results such as the probability of each class. Since this soft label gives more information to the student network compared with conventional label data, it helps to make a further smaller network through knowledge distillation. Smaller models have a lower evaluation cost and can therefore be used on less powerful hardware such as mobile devices (Fig. 2.11).

2.1.5 *Learning According to Run-Time Data Collection: Active Learning vs. Online Learning*

Active Learning

Active learning focuses on finding ways to label the smallest number of samples while achieving the greatest performance gain. To minimize labeling costs while maintaining performance, the most valuable samples are selectively selected from unlabeled datasets. Active learning asks the labels to an oracle, and then, the supervisor such as a human gives proper labels to DNN. Active learning is similar to semi-supervised learning, which is also a weakly supervised learning technology, in that it handles both unlabeled and labeled data, but does it differently (semi-supervised learning does not use oracles).

Online Learning

Online learning is a machine learning technique that updates the model as new data are collected or arrive sequentially, as opposed to batch learning (or offline learning), which first collects a dataset from multiple instances and then trains the model once. You can update the model later when you receive new data in run-time. Online learning is currently a common way to train DNNs to adapt to a new environment. The key concept of online learning is obtaining labels in run-time. The label-obtaining method can be varied, and it can also utilize the unsupervised or semi-supervised manner. However, it is different from active learning because it does not depend on help from the oracle. It is also a kind of incremental learning that tries to solve this biased inference problem (Fig. 2.12).

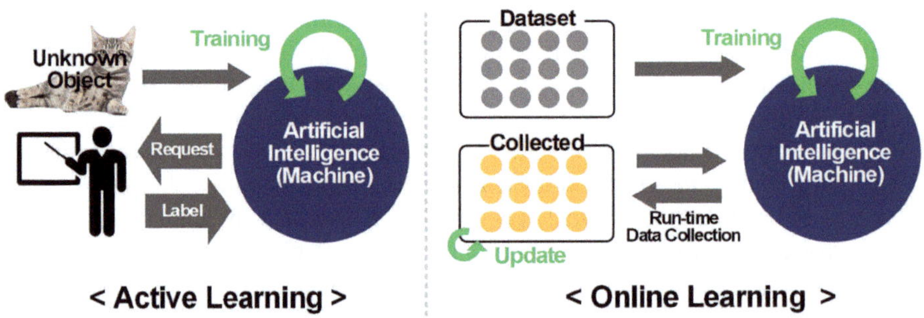

Fig. 2.12 Overview of active learning and online learning (dark blue: shared layer among various tasks, light blue: independent layer for each task, red box: target layer for current task)

2.1.6 *Multi-task Learning*

Fine-Tuning

Fine-tuning describes a technique for updating the training from previously taught model weights and modifying the architecture for new use. The amount of the new dataset we have and how similar it is to the dataset the pre-trained model trained on determine the best approach:

1. Dataset for pre-training is large, and the similarity between the previous and target datasets is small: You need to train the entire layers of DNN. Training is possible because the dataset is large. Even if there is little overlap between the datasets, re-training is recommended because the structure and parameters of the model can still be used.
2. Dataset for pre-training is large, and the similarity between the previous and target datasets is large: You need to train the last layers of the classifier and the convolution basis. It is the best way to minimize the re-training cost, but training both backbone convolutional layers and fully connected layer-based classifier shows the best training accuracy compared with the classifier-only training.
3. Dataset for pre-training is small, and the similarity between the previous and target datasets is small: This is the most challenging scenario and usually trains a part of the classifier and backbone convolutional layers. A small number of fine-tuned layers have little effect because they have very little data. Thus, it can suffer from an overfitting problem when there are a large number of fine-tuned layers. Therefore, it is important to accurately estimate the size of backbone convolutional layers during the training from scratch.
4. Dataset for pre-training is small, and the similarity between the previous and target datasets is large: There is only one trained classifier. Fine-tuning multiple layers leads to overfitting because there are less data. As a result, you can use the original part of the classifier's feature extraction and fine-tune only the FC layer.

Joint Training

Joint training involves both old and new tasks and is called joint training. This is the most effective approach, and it assumes that all data have been compromised. It takes a lot of time, but performance will be guaranteed.

Learning Without Forgetting

Learning without forgetting [148] fine-tunes the weights with new data but also improves the classification results of existing data. It uses only examples for the new tasks (old task's examples: x), and it should remember the original DNN's response of old tasks (Figs. 2.13).

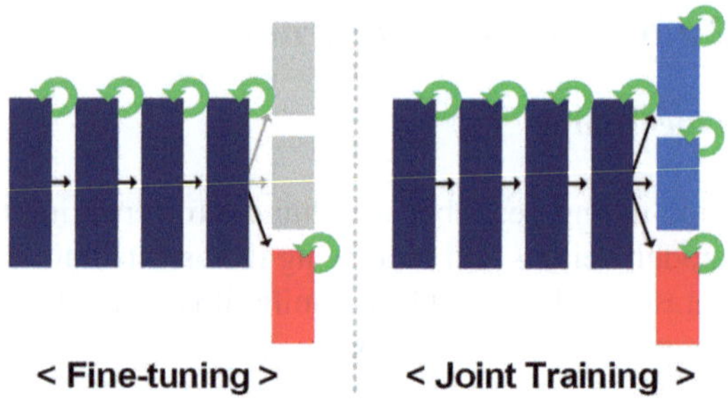

Fig. 2.13 Overview of fine-tuning and joint training

2.2 Universal Approximation Theorems

We can have a general question, "Can a DNN learn any tasks?" Mathematically speaking, the purpose of the DNN is to find any mathematical function, $f(x)$. $f(x)$ is the function that maps attributes (x) to output (y) with its corresponding weights or parameters (w), but its complexity is not predictable, which means $f(x)$ can be arbitrarily complex. Then, can DNN have the functionality of $f(x)$ even though $f(x)$ is super complex?

At the initial stage of the AI research, people tried to solve this problem with the perceptron that consists of only one single layer, which directly connects inputs and outputs with weights. It computes linear computation and then directly transfers to a non-linear activation function to get output. This concept was similar to the human brain's neuron, but their hope was broken after facing the XOR problem. In 1989, George Cybenko [147] proved that "Finite linear combinations of compositions of a fixed, uni-variate function and a set of affine functions can uniformly approximate any continuous function of n real variables with support in the unit hypercube." In other words, whatever $f(x)$ is, it can be well approximated even with the DNN that has only one hidden layer. Although he proved only one hidden layer is sufficient to approximate any continuous function, but it did not restrict the number of neurons. According to his proof, we cannot estimate the required number of neurons before training, and also, it did not tell us that the parameters of the single hidden layer DNN can be trained well. After his theory, the XOR problem was solved easily, but the optimal training solution is still not suggested and challenging.

2.3 Perceptron

2.3.1 Basic Structure of Perceptron

Figure 2.14 shows the structure of the most basic perceptron. Our goal is to find the weights that best predict the output for the input. To do so, we find the weights by repeating the process of finding the output using the training set (feed-forward),

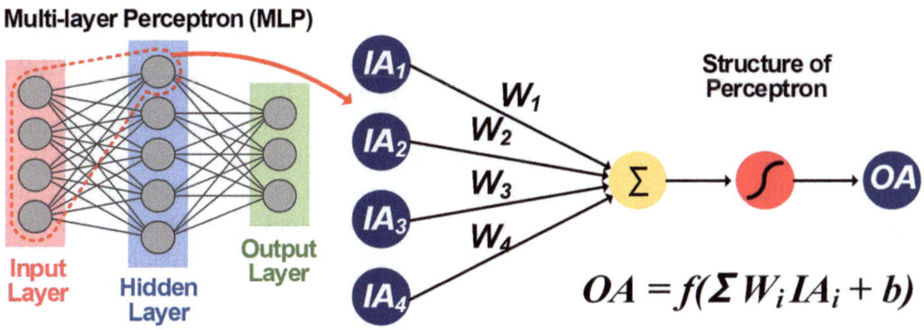

Fig. 2.14 Architecture of multi-layer perceptron and a single perceptron

finding the error for the output, and then finding the direction in which the error decreases and updating (such as using back-propagation). A single neuron, called perceptron, consists of linear computations (matrix multiplication) and non-linear activation function.

2.3.2 *Non-linear Activation Function*

Artificial neural networks started with the basic concept of a single perceptron, and the image above shows how a neural network transmits signals. Several signals are coming in and transmitting the output. At this time, it is the activation function that determines the strength (strength) of the transmitted output signal.

As the word activation suggests, it is the role of determining whether a combination of input signals triggers the activation of a neuron. And when studying the activation function, non-linearity is always mentioned, and that is why it is so important. If so, I wonder why non-linearity is important in deep learning, and I think it can be simply explained as follows. As the data become more complex and the dimensions of features increase, the data distribution appears non-linear and non-linear, and it is impossible to express it with a simple linear boundary, so a non-linear boundary is required.

2.3.3 *Multi-layer Perceptron (MLP)*

Consider the case of classifying a dog or cat using black or white images. Consider a case where increasing the value of each pixel increases or decreases the probability of determining that it is a dog. This is not reasonable. Because, if this happens, it means that in the end the puppies are all black and the cats are all white or vice versa.

To figure out what is in an image, you have to consider the very complex relationships between inputs and outputs, and the possibility that patterns can be specified through relationships between many features. In this case, the accuracy of the linear model will be low. We can use more than one hidden layer together to build a model using more general functions. The easiest way to implement this is to stack other layers on top of each layer. The output of each layer is fed to the input of the layer above it, which repeats until the last output layer. This architecture is commonly referred to as a "multilayer perceptron." In other words, MLP builds multiple layers in succession.

For example, Fig. 2.14 shows a multi-layer perceptron that has 4 inputs and 3 outputs, and the hidden layer in the middle has 5 hidden units. Neurons in the hidden layer are all connected to the inputs in the input layer. Neurons in the output layer and neurons in the hidden layer are also connected. Therefore, both the hidden layer and the output layer of this multi-layer perceptron are fully connected layers.

2.4 DNN Training with Back-Propagation

The goal of our neural network is to predict well as close to the true value as possible based on logistic regression, which is also normalized. To do this, the neural network must be trained, and in the end, training means minimizing the loss, which is the error between the actual value and the predicted value. It usually minimizes the loss by using the gradient descent algorithm (the typical method is back-propagation) we are familiar with in the neural network. A unit in a neural network is multiplied by an input x and a weight W to create $g(x) = Wx$. Put this in the sigmoid function h, and send the derived value $y = f_W(x) = h(g(x))$ to the next layer. The end goal is to find the weight W of all units in the neural network that minimizes the loss. This is expressed as optimizing a neural network.

2.4.1 *Forward and Backward Paths of the Back-Propagation*

Forward propagation means calculating and storing variables in order from the input layer to the output layer of the neural network model. We call it "inference." Backward propagation refers to a method of calculating gradients for parameters of a neural network. In general, back-propagation calculates and stores the gradients of intermediate variables and parameters of the objective function associated with each layer of the neural network in order from the output layer to the input layer. When training a network, forward propagation and backward propagation are mutually dependent relationships (Figs. 2.15, 2.16, 2.17, and 2.18).

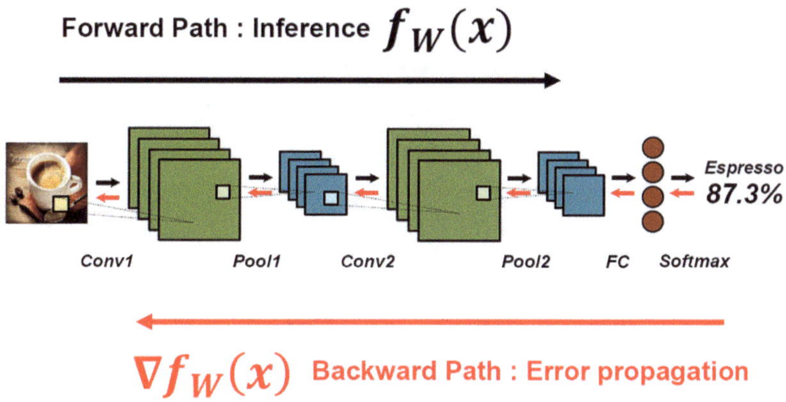

Fig. 2.15 Forward and backward paths of the back-propagation

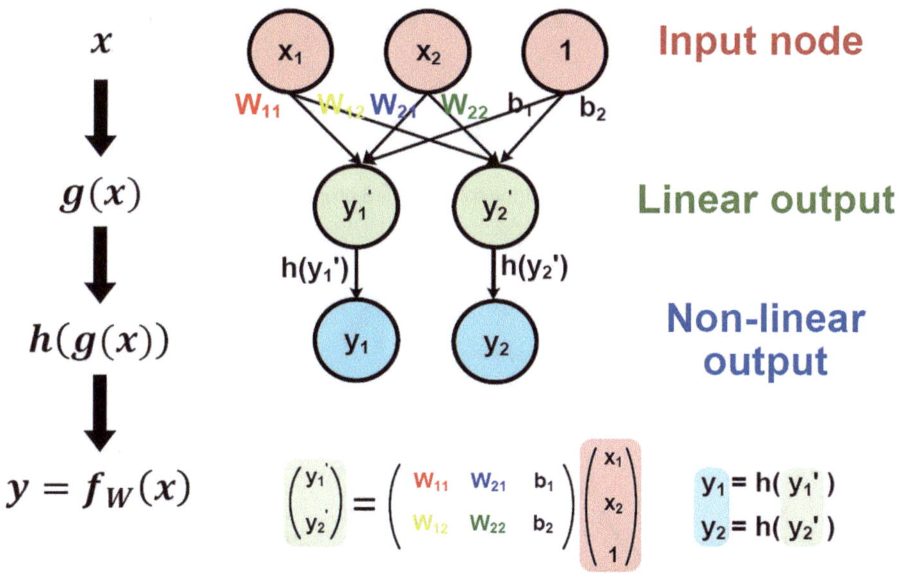

Fig. 2.16 Mathematical representation for forward propagation

2.4.2 Gradient Descent

From multi-variable calculus, we know that the gradient of a function, γf, at a specific point will be a vector tangential to the surface pointing in the direction where the function increases most rapidly. Conversely, the negative gradient $-\gamma f$ will point in the direction where the function decreases most rapidly. This fact we

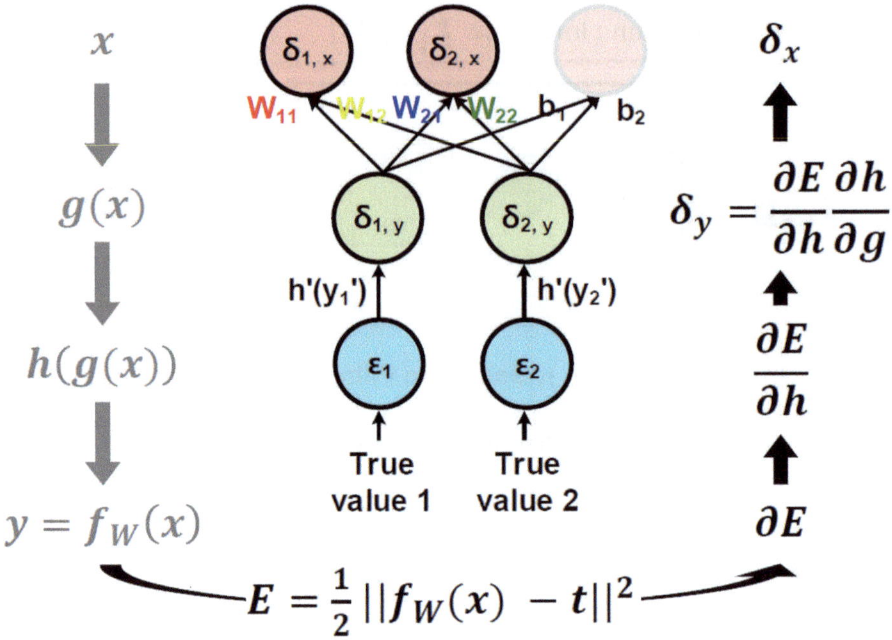

Fig. 2.17 Brief mathematical representation for backward propagation after loss (E) calculation

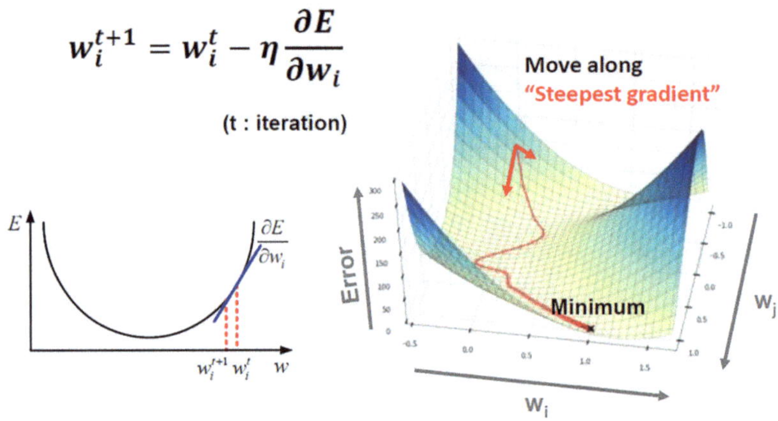

Fig. 2.18 Overview of gradient descent

can use to calculate new weights W_i^{t+1} from our current weights W_i^t. The gradient may be calculated in two different ways: (1) numerical gradient, which is slow, approximate, and more error-prone, (2) analytic gradient, which is simple, rapid, and accurate based on calculus.

$$\frac{df(x)}{dx} = \lim_{h \to 0} \frac{f(x+h) - f(x)}{h}$$

Numerical gradient: slow :(, approximate :(, easy to write :)
Analytic gradient: fast :), exact :), error-prone :(

Fig. 2.19 Two types of gradient descent: (1) numerical gradient and (2) analytic gradient (Source: Stanford University CS231n lecture)

Numerical Gradient Descent

Numerical gradient descent is the intuitive methodology to realize gradient descent by using the definition of the gradient. With the help of the slope formula, we can calculate the gradient numerically as shown in Fig. 2.19. This calculation iterates over all dimensions one at a time, which makes a minor modification h along each dimension and then calculates the partial derivative of the loss function along each dimension by seeing how much the function changed, using the gradient formula we provided previously. However, this slope calculation should be performed for every weight element. Since a single slope calculation requires two independent DNN inferences, it requires a total of 16 inferences if the number of input and output neurons is 4. The required computation amount can be increased significantly when the network consists of a large number of neurons or has deeper layers. Finally, numerical gradient descent is not considered for DNN training nowadays due to its heavy computation amount.

Analytic Gradient Descent

Analytic gradient descent actually makes it possible to find partial derivatives directly using differential formulas. Back-propagation is directly related to this analytic gradient descent, which computes a graph along the chain rule and computes all variables in that path. To apply analytic gradient descent, we should know the mathematical expressions of the partial derivatives of the entire layers. Then, the slope of the loss space can be easily calculated with simple multiplication among partial derivatives values. Normally, the values of partial derivatives depend on their input and output values. To obtain a partial derivative, we should remember the intermediate input and outputs of the network. This is one of the reasons why back-propagation uses much more memory than performing simple predictions. In other words, to apply the chain rule, all intermediate variables must be stored, so that gradient tensors can be calculated. Unlike inference-only cases, DNN training needs to store all intermediate features and uses them to calculate partial derivatives. When the batch gradient descent is applied, it requires more memory to store more intermediate activations caused by batch processing.

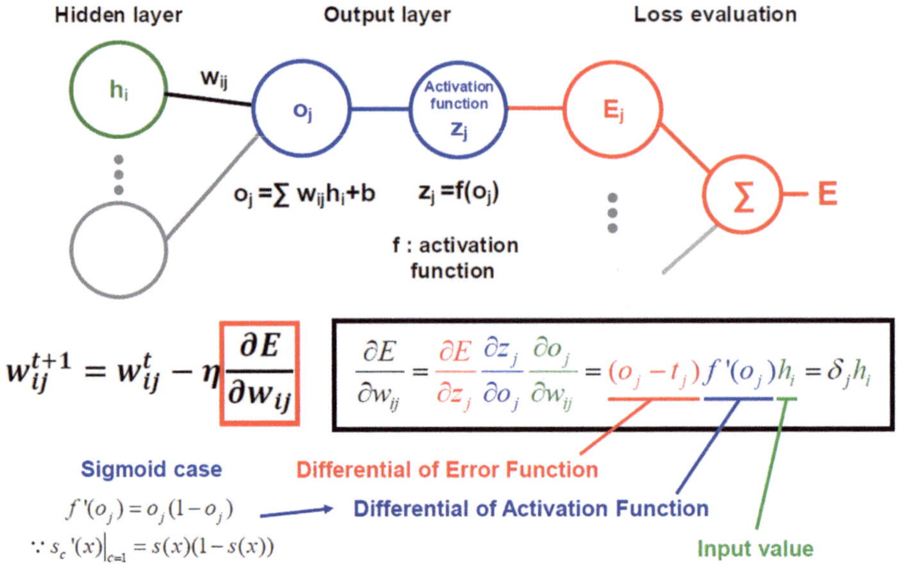

Fig. 2.20 Mathematical explanation for back-propagation of the last layer

Figures 2.20 and 2.21 show typical back-propagation example to understand the calculation intuitively. We summarize the mathematical expressions required for the last layer training. We assume that the loss E is calculated by L2 distance $E = \frac{1}{2}\Sigma(o_j - t_j)^2$ between inference results o_j and corresponding labels t_j. As shown in Fig. 2.20, final weight gradient can be calculated by multiplying propagated error and input activation. It is the same as hidden layer. As depicted in Fig. 2.21, only the difference is that multiplied error reflects its propagation path. If a neuron is connected to multiple neurons, propagated errors from the entire neurons should be aggregated. The final aggregated error is used for calculation of weight gradient.

As summarized in Fig. 2.22, back-propagation is based on analytic gradient descent; thus, it should remember input activations calculated during the feed-forward stage, inference. Furthermore, it performs error propagation to propagate loss from the last layer to entire prior layers. After calculation of entire errors, memorized input activations and errors will be multiplied to make final weight gradients.

2.4.3 Back-Propagation in Convolutional Neural Network

In Sect. 2.4.2, we explained mathematical expression and analysis of back-propagation in multi-layer perceptron. Back-propagation in the convolutional neural network is not that different. As understood in Sect. 2.4.2, error propagation is

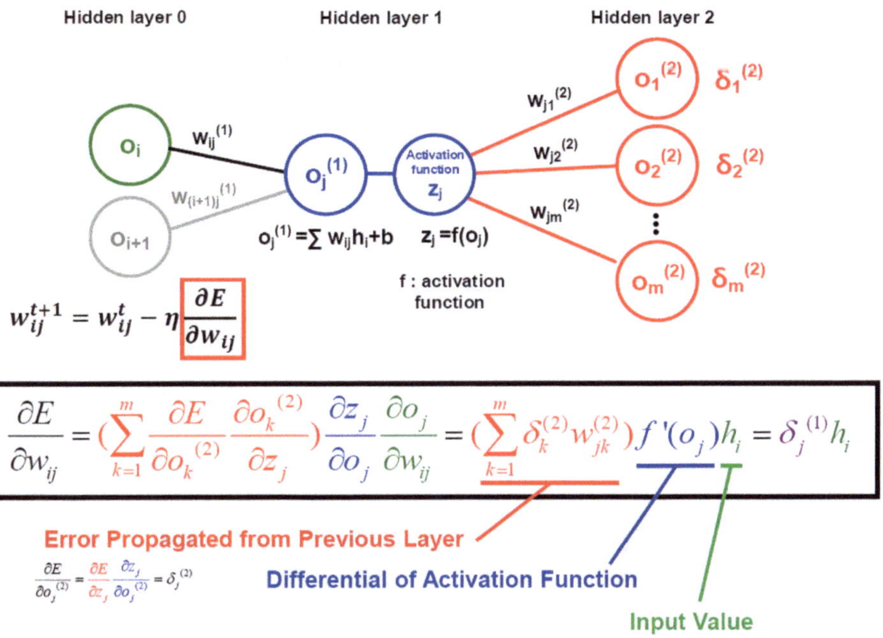

Fig. 2.21 Mathematical explanation for back-propagation of the hidden layer

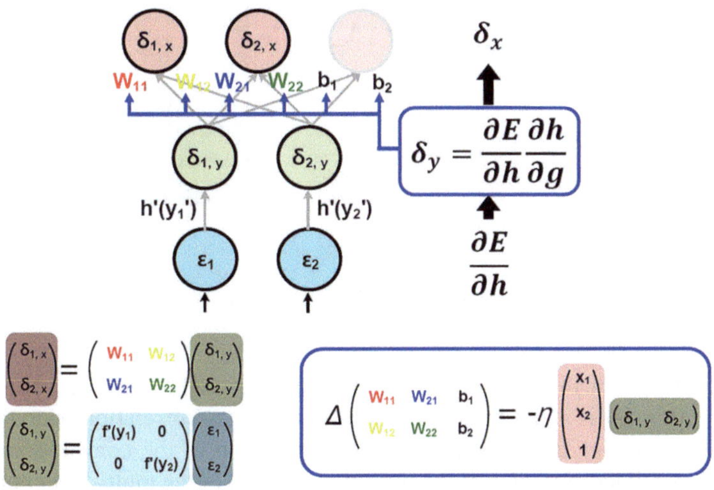

Fig. 2.22 Overview of required computations in weight-gradient-update stage

performed by considering connections among neurons. It is directly reflected in back-propagation-related equations. As shown in Fig. 2.23, convolutional neural network can be mapped to multi-layer perceptron. In the real implementation of deep learning, the convolution operation is usually substituted with cross-correlation

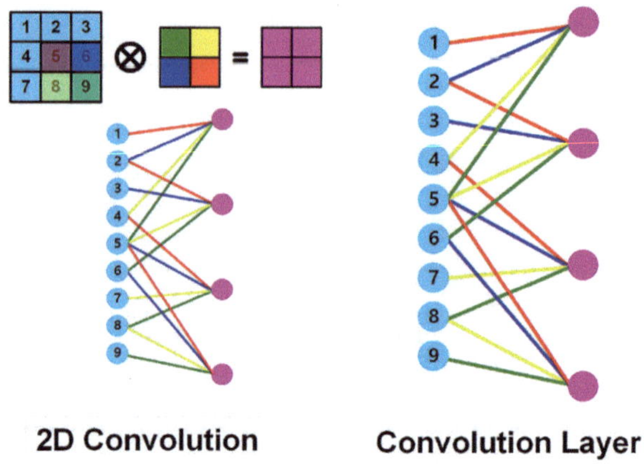

Fig. 2.23 Mapping the convolutional layer to multi-layer perceptron structure

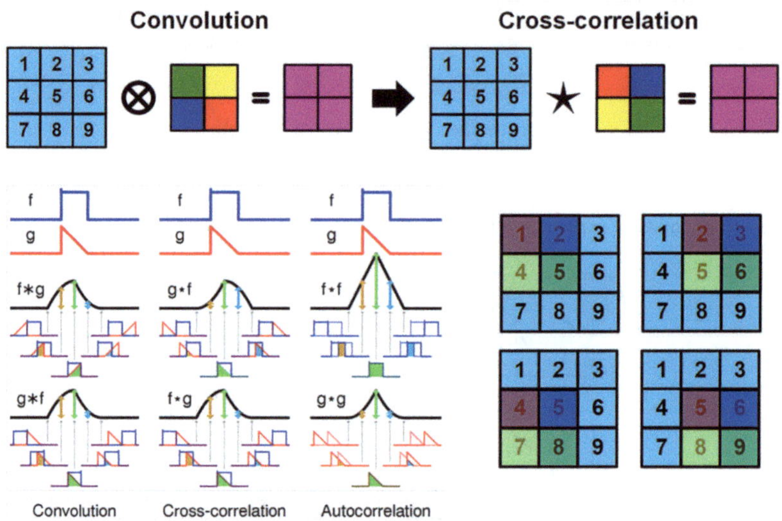

Fig. 2.24 Computational relationship between convolution and cross-correlation

operation as shown in Fig. 2.24. From now on, the convolution operation that is referred in this thesis is not that different from the cross-correlation operation.

As shown in Figs. 2.23 and 2.24, back-propagation of the convolutional neural network is not that different with the back-propagation of the multi-layer perceptron. Instead, since the convolutional neural network shows a sparse connection, we can simplify the computation required for back-propagation that can be reduced by using convolution instead of matrix multiplication. Figures 2.25 and 2.26 summarize the back-propagation of the convolutional neural network. There are three key observations: (1) if there is convolution during the forward propagation, it also

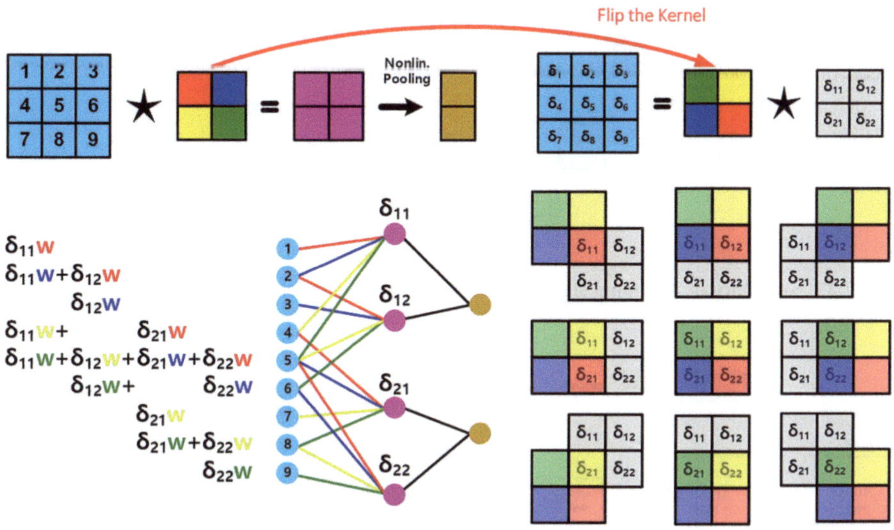

Fig. 2.25 Back-propagation-based error propagation of convolutional layer

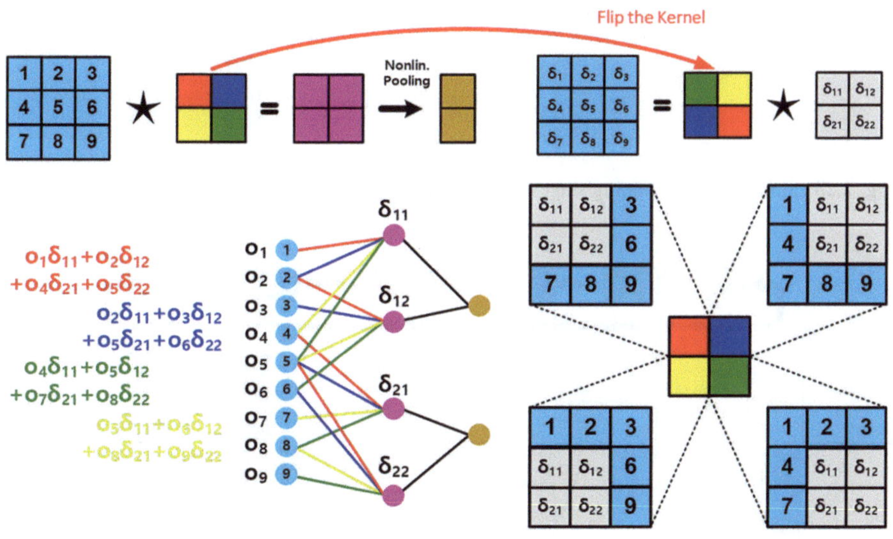

Fig. 2.26 Weight-gradient-update of convolutional layer

needs convolution during the backward propagation. (2) The weight kernel should be rotated 180° during the backward propagation. (3) Input activation and output error map should be computed with convolution operation to calculate the gradient (Figs. 2.27, 2.28, 2.29, and 2.30).

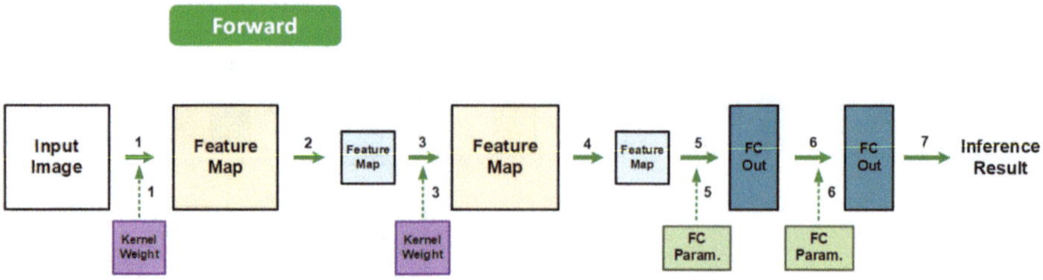

Fig. 2.27 Summary of 3 training stages of convolutional neural network (1) feed-forward

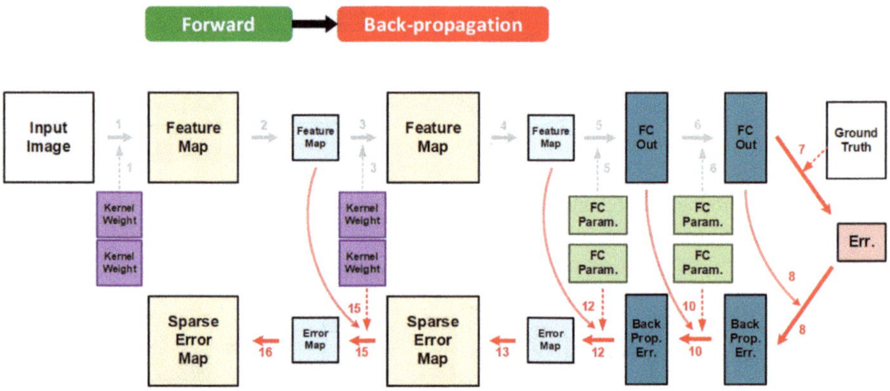

Fig. 2.28 Summary of 3 training stages of convolutional neural network (2) back-propagation

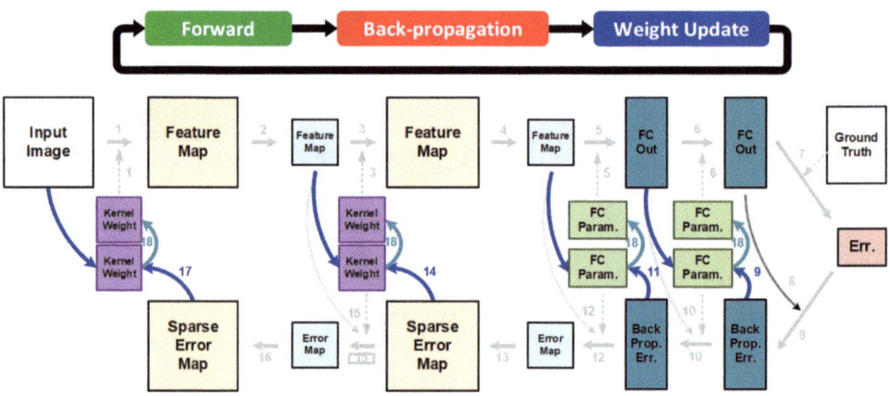

Fig. 2.29 Summary of 3 training stages of convolutional neural network (3) weight-gradient-update

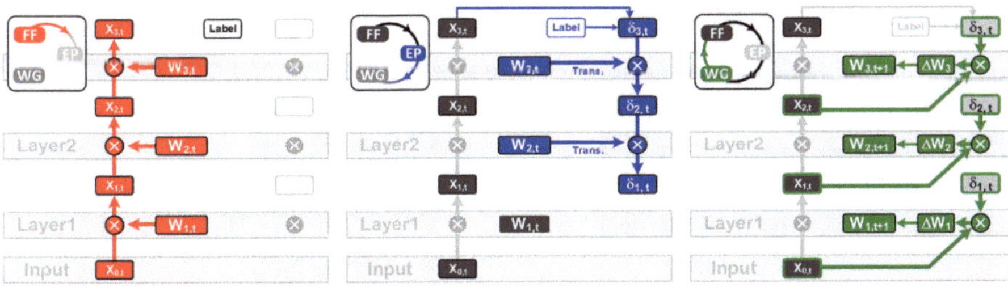

Fig. 2.30 Overview of back-propagation-based deep neural network training

2.5 Brief Summary of Required Computations During BP

Even though there are many DNN training methodologies, back-propagation (BP, [40]) is the mostly used DNN training method. Thus, we will briefly explain distinct computing characteristics of the BP compared with the inference. We will also introduce the main challenges to realize fast and energy-efficient training at the micro-AI systems.

2.5.1 Back-Propagation of Fully Connected Layer and Convolution Layer

BP is a loss minimization method based on gradient descent by using the chain rule. It stores all intermediate activations and errors generated in every layer and loads them to calculate the new gradients. The BP based training can calculate accurate gradients of the multiple parameters with just one inference. Therefore, BP can significantly reduce the computational requirement, but it needs large storage to remember entire activations and errors to update the weights. As shown in Fig. 2.31, the BP generally consists of three training stages: (1) feed-forward (FF), (2) error propagation (EP), and (3) weight-gradient-update computing (WG). In the FF stage, DNN gets inference results with the pre-trained weights. After that, the loss function is calculated and generates the error. In the EP stage, this error is now propagated from the last layer to prior layers step by step. Lastly, the gradient of the weight is calculated by aggregating activations and errors. This training process is general for any type of DNN, but the required computing methods can be varied according to the target layer type. In this section, we will explain the detailed computation of the three training stages, taking two commonly used layer types, the fully connected layer (FCL) and convolution layer (CL) as examples.

Back-Propagation of Fully Connected Layer

When input and output neurons are fully connected in a layer, we call it FCL, and it is the most general layer type of the DNN. For example, multi-layer

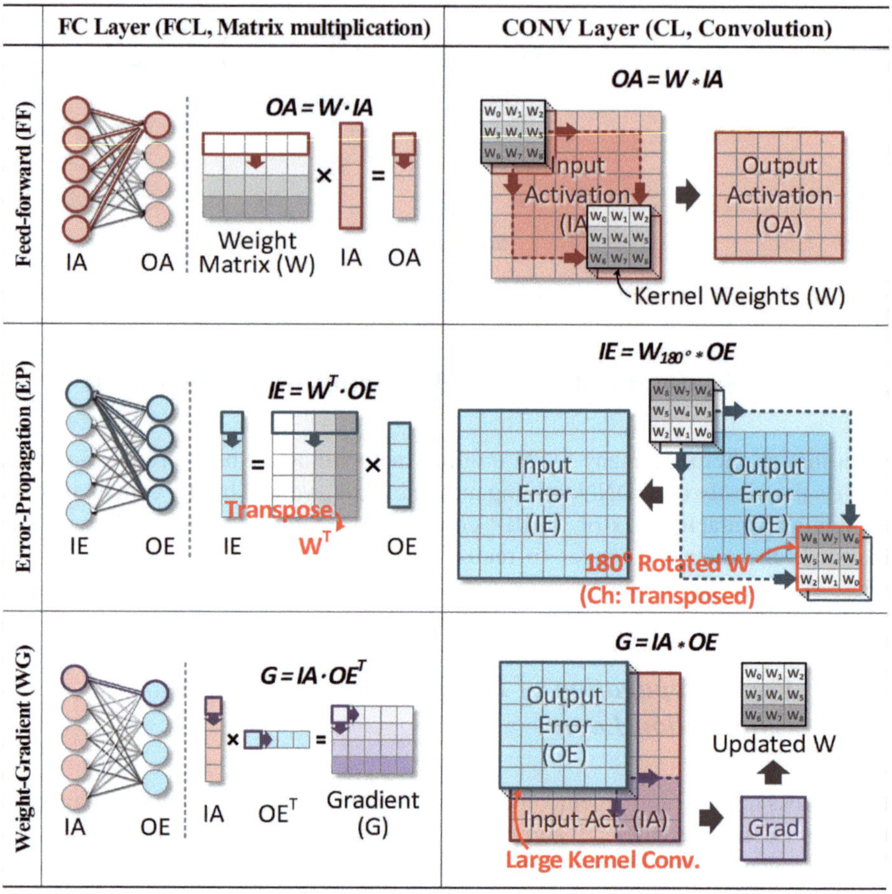

Fig. 2.31 Brief summary of back-propagation-based DNN training

perceptron and recurrent neural network use the stacks of FCLs to complete the network architecture. Convolutional neural network (CNN) also utilizes FCLs for classification or channel attention. The FF stage of FCL is based on matrix multiplication of input activation (IA) and weight. In the EP stage, the computation method is similar, but the input operands are replaced with error and transposed weight. After both FF and EP stages, IA and output error (OE) are multiplied to create the gradient of the weight. The WG stage needs element-wise multiplication in the single-batch gradient descent.

Back-Propagation of Convolutional Layer

Although the FCL needs matrix multiplication and element-wise multiplication for training, CL only needs convolution operation. Instead, the shapes of the weight kernel are varied according to the training stages. It needs a 180° rotated kernel to propagate the errors. The position of the input and output channels is also switched. As similar to FCL, CL also needs the transpose-read due to this channel switching. After the EP stage, the gradient of the weight is calculated by the convolution of IA and OE. Unlike the FF and EP stages, the WG stage needs large-kernel-size convolution because the size of OE is as large as the IA. As a result, the distance of the spatial data-reusing at the WG stage is not as long as the other training stages.

2.6 Challenges of BP at the Micro-AI Systems

The DNN training requires not only FF stage but also EP and WG stages. Additionally, DNN training is sensitive to accuracy degradation so that optimization methods of inference processors may be useless because they sometimes degrade its training performance. This difference induces new challenges to design on-chip training processor, and they are summarized as follows.

2.6.1 Reading of Transposed Weight

The EP stage reads transposed form of the forward weight (FW) before it computes the convolution. The CL requires a 180° rotated kernel as well as a transposed kernel. Reading of 180° rotated data can be realized simply by reading data in reversed order. However, reading of transposed weight is not that simple because it reads discontinuous data from the external memory where the weight is stored. Discontinuous data access pattern caused by transposed weight disturbs burst-mode read of DRAM and induces slow and inefficient DNN computing.

2.6.2 Maximizing Throughput Even with the Limited Hardware Resources

DNN training requires a much larger amount of computations compared with inference. It needs to perform additional two training stages and repeats this training loop until the loss curve is converged. It also utilizes batch gradient descent that uses multiple inference results to generate a single gradient.

The required batch size may be varied according to its target training applications. Generally, the batch size becomes relatively small if the target network requires large memory size such as image-to-image translation networks. On the contrary, as shown in Fig. 2.32, deep reinforcement learning (DRL) networks utilize large batch sizes because small size is relatively small. In the object classification, its training performance can be degraded significantly with the batch size < 4, but beyond 4, any significant difference in training performance is found (Fig. 2.33).

Batch Size	1	1	4	32	40-100	256	2048
Papers	[24, 32]	[26]	[21]	[34]	[23]	[30]	[20]
DNN Type	CNN	CNN	CNN	CNN	CNN	MLP	MLP
Usage	GAN (Img2img)	Classification	Classification	Classification*	Classification	DRL	DRL

*On-chip training of classification DNN to realize tracking

Fig. 2.32 Typical batch size used in DNN training processors

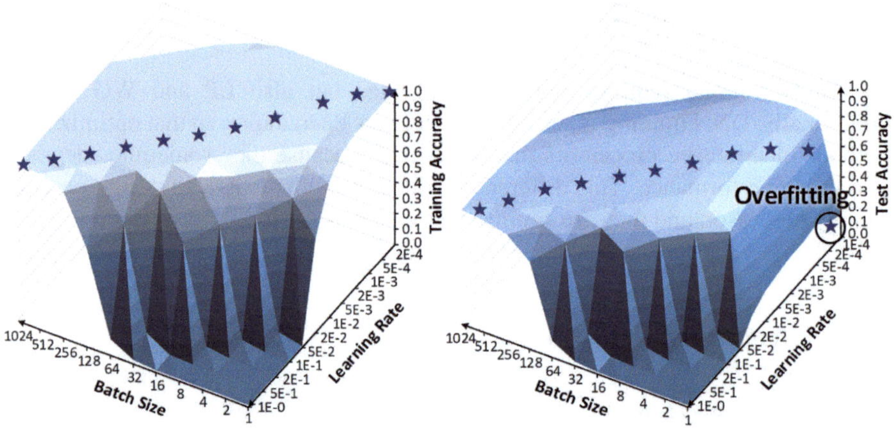

Fig. 2.33 Training and test accuracy of CIFAR-100 according to batch size and learning rate. (Blue star = highest accuracy among various learning rates) batch size and its best learning rate show a positive linear correlation

For this reason, the conventional GP-GPU utilized large batch parallelism to reduce the overall training time (Fig. 2.34). However, it is impossible in mobile training processors because of their limited on-chip memory size. Therefore, a novel throughput improvement method is necessary for the training processors to accelerate training even with the small-batch parallelism.

2.6.3 High-Bit-Precision Requirement

During the inference, the final decision of the DNN is mainly affected by the large values of the IA of which the distribution is much narrower than the training [29]. For this reason, it is easy to decrease the bit-width during the inference acceleration. Unlike the inference, it is not so simple in the training because even small values of IA may include important information for loss minimization. The naïve quantization methods can destroy this information and slow down the training. Furthermore, the errors computed in the EP stage have a wider distribution compared with the IA, which is insufficient to be represented with low bit-precision. As a result, training

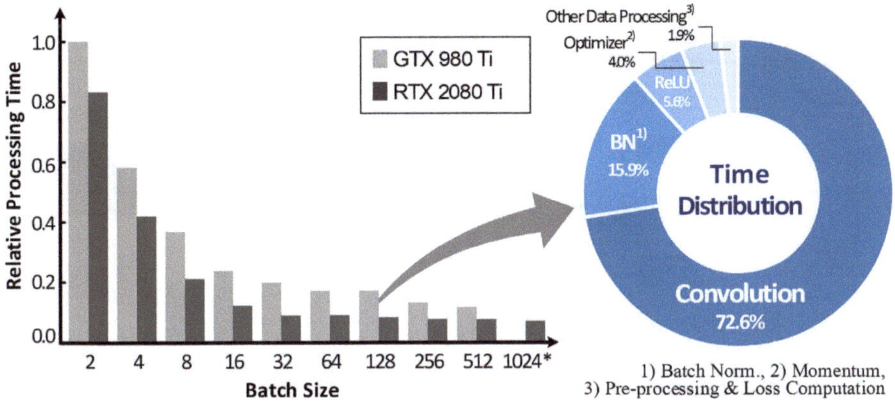

Fig. 2.34 Analysis of DNN training at the GP-GPU (ResNet @ CIFAR-100)

requires high-bit-precision representation, and this requirement becomes the main obstacle to realize training at the micro-AI systems because it totally decreases the efficiency of the processor.

2.6.4 Memory-Intensive Operations During Weight-Gradient-Update

As described in Fig. 2.32, intermediate IA and OE should be stored until the weight gradient is calculated. In addition, although data-reusing can be partially done in the CL, its reusing amount is much smaller than in the other training stages. Computing in the WG stage becomes similar to matrix multiplication that appears in the FCL. After gradient calculation, it additionally loads the weight and momentum to finalize the weight update process. It causes high arithmetic intensity (the number of operations per byte) so that its performance highly depends on its external and internal memory bandwidth. Consequently, the reduction of memory access during the WG is the most important optimization method to accelerate its computation.

2.6.5 Backward Locking Problem

The BP propagates errors sequentially from the last layer to the first layers as depicted in Fig. 2.35. Therefore, the weight of the first layer can be updated only when the error propagation of the entire layers is completed. In other words, FF, EP, and WG stages should be computed sequentially and cannot be processed in parallel. It is called the backward locking problems [34], and this sequential process

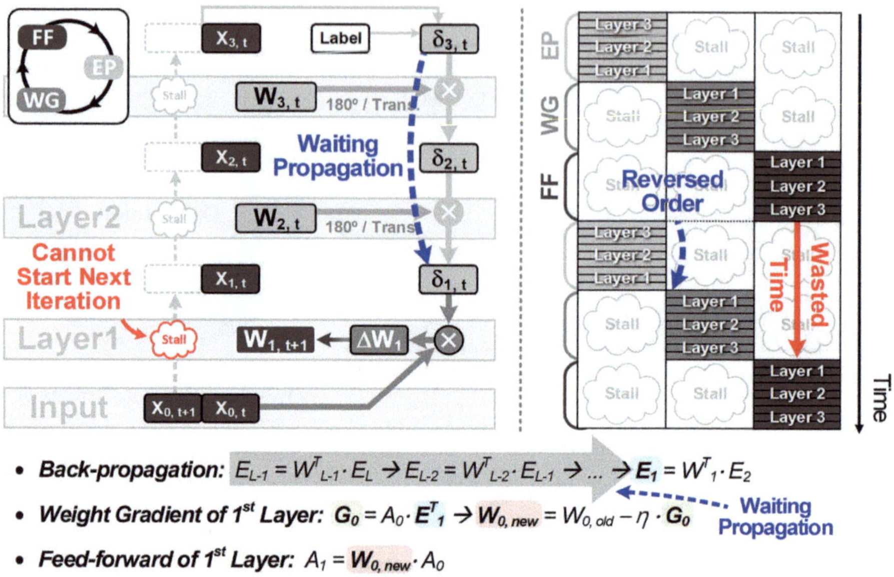

Fig. 2.35 Low training speed caused by backward locking problem

has two disadvantages. First, three training stages require the same weights, but the weights should be reloaded every training stage. Second, it cannot hide the latency of the EP and WG stages to significantly reduce the framerate during online DNN training such as object tracking [41] or temporal knowledge distillation (TKD, [42]).

2.7 Biological Plausibility of Back-Propagation

Back-propagation is algorithmically superior, but brain, cognitive, and neuroscientific research has shown that it is not physiologically valid:

1. **Synapse in brain is not unidirectional**: A single neuron's impulse transmission in a real brain's genuine neural network is unidirectional. This is the most fundamental problem with how neurons are built is that the synapses that connect neurons to real neurons only flow in one direction, making back-propagation and other two-way communication impossible. Then, is feedback itself impossible? The unidirectional transmission pathway of a single neuron does not rule out the possibility of a return pathway. In fact, there are many pathways that allow feedback in neural connectivity structures. For example, there is a correlation. Feedback is only possible if there is a relationship. Thus, the brain can back-propagate through this feedback channel. However, there are still some issues with the viability of back-propagation in the real brain.

2. **Weight Transport Problem**: The weight of feedback synaptic connections from layer $n + 1$ to layer n cannot equal the weight of synaptic connections from layer $n + 1$ to layer n in a real brain's genuine neural network. We call it a "Weight transport problem." The connection weight data utilized by the forward propagation of the $n + 1$ layer are needed in the back-propagation method to update the forward propagation route weights of the n layer. As a result, the weights of the $n + 1$ layer must be exactly equal to the weights of the feedback channels from the $n + 1$ layer to the n layers (transposed read during error propagation as mentioned in Sect. 2.6.1). The weights of the forward propagation path are modified for each forward propagation and backward propagation. In fact, in real neural networks, the weights of the feedback paths for n layers are always exactly symmetrical with the weights of the forward propagation paths at layers $n + 1$. It is almost impossible in the brain.
3. **Local Learning**: The parallel structure of a real brain's genuine neural network is made up of little connections between nearby neurons. Global error information transmission from the output layer to non-adjacent layers such as MLP is not conceivable in this arrangement. Global error transmission is not possible in a real brain. Real numbers from the output layer propagate backward and propagate to the input layer away from the MLP's back-propagation (this is called global learning). In other words, we cannot build complex feedback channels from the source of the mistake to the distant neurons. To sum up, neural networks in the brain perform local learning, not the global learning adopted in back-propagation.

Clearly, a principle of back-propagation is entirely different from the training methodologies adopted in a genuine brain. For this reason, many researchers commonly have the same idea, "Back-propagation is not the complete, final form of AI training." Of course, on-device training processors in current ages focus on the acceleration of back-propagation-based training. However, the other alternatives should be co-researched to offer fast, efficient, and adaptive training. Sometimes, the solutions can be found in the real brain's training methodologies. There were several brain-inspired DNN training methodologies studied in previous research. In this section, we will first introduce previous AI training methodologies researched in the past. And then, we introduce non-back-propagation learning methodologies such as local learning, and semi-back-propagation learning methodologies such as feedback alignment will be explained.

2.8 DNN Training Without Back-Propagation

2.8.1 Genetic Algorithm

Developed by John Holland in 1975, genetic algorithms [149] focus on the ability to live organisms to adapt to their environment by engineering genetics and evolutionary processes in the natural world. It is an optimization technique based on the principles of natural selection, first discussed in "Adaptation to natural and artificial systems." The application of genetic algorithms in search, optimization, and machine learning is widespread.

By describing a potential answer to a problem in a fixed-form data structure and making incremental changes, genetic algorithms generate superior solutions. In other words, chromosomes convey a potential solution to an immediate problem, and by gradually changing that potential solution, a better solution is subsequently created. All potential answers are considered organisms or individuals, and the set of them is called a population. One or more chromosomes usually make up an individual, and the genetic operators that change chromosomes are known per se. There are three basic operators: In the next step, the survival distribution of the crossed individuals is determined by selection according to the adaptive distribution among the populations. That's because they are based on an adaptive distribution, crossover (the process of changing genes on two chromosomes to create a new individual), mutation, and increased adaptiveness (forcing the values of certain parts of genes to change).

Evolution Strategy

A wind tunnel experiment was conducted in 1963 by two Berlin Technical University (TUB) students to determine the shape of the item with the least air resistance. One of the two students, Ingo Rechenberg, had the idea to randomly change the shape-defining elements, such as mutations after the search for gradient methods failed. At this point, an evolutionary strategy emerged. Reckenberg proposed two individual evolutionary techniques in 1973 and extended them to include more than two.

Another creator of evolutionary strategies, Hans Paul Schwefel wrote a paper in 1977 entitled Evolutionary Strategies and Numerical Optimization. Because the objective function could not be generated in an analytic form, traditional optimization approaches could not be used to solve technical optimization problems. The pre-existing normally distributed mutations were given to one parent in every generation of the original $(1 + 1)$ evolutionary strategy of two persons and replaced the parent when superior children appeared (Fig. 2.36).

For the initial evolutionary strategy with this basic structure, the management of the amount of variation and the convergence rate and other theoretical studies were conducted. This study found that one-fifth of all mutations resulted in effective mutations. The 1/5 success rule is another name for this. The $(\mu + \lambda)$ evolutionary approach, which can incorporate recombination processes using parental organisms, was created from the first evolutionary strategies that rely solely on mutations. $(1 + 1)$ evolutionary technology is the same as a mutant system and is a way of regulating growth. Schwepel divided multi-individual evolutionary techniques into $(\mu + \lambda)$ and (μ, λ) evolutionary strategies, where the population size is μ and the number of children born in each generation is λ.

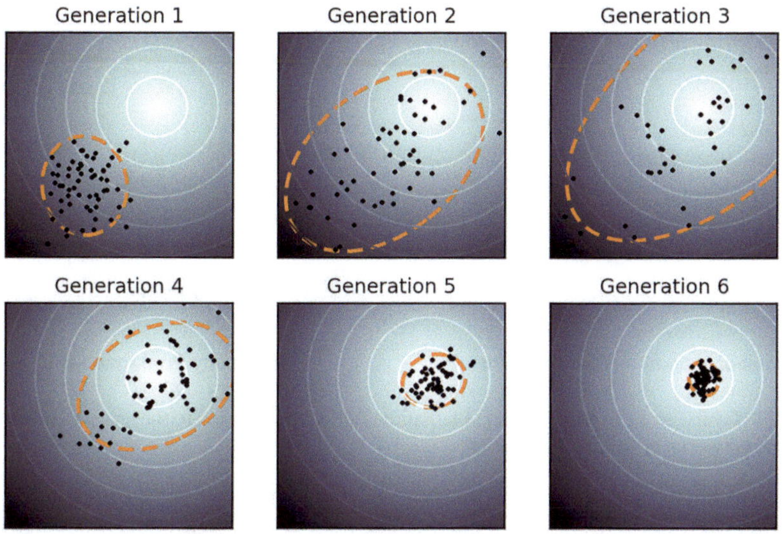

Fig. 2.36 Custom image displaying the directional optimization of the evolution strategy algorithm. (Source : https://en.wikipedia.org/wiki/CMA-ES)

This generalized evolutionary technique was able to replicate the natural evolutionary processes of random reproduction, mutation, and selection by adding the idea of groups of people. In many test cases, evolutionary solutions outperformed other approaches. Evolutionary techniques can be used to solve almost any type of optimization problem because they require no background knowledge of the problem other than an individual's method of evaluation. Higher-order, multi-modal, non-linear problems with both linear and non-linear constraints can be solved using evolutionary techniques. The simulation result can be utilized as an objective function. You do not have to have an analytic form.

2.8.2 *Local Learning*

Unlike biological neurons that learn from local information, back-propagation allows for the learning of individual nodes while requiring global knowledge of the entire node. Artificial neural networks have recently experimented with models that can be trained locally, just like biological neurons, to overcome the biological de-approximation of back-propagation.

Because local models are solely trained on points relevant to particular test data, local learning algorithms frequently perform better than global learning algorithms. The local learning approach has demonstrated good performance, which suggests that labels for the locations may be accurately approximated based on their neighbors. This local learning is motivated by unsupervised learning because it wants to make a well-trained model without manual supervision from oracle.

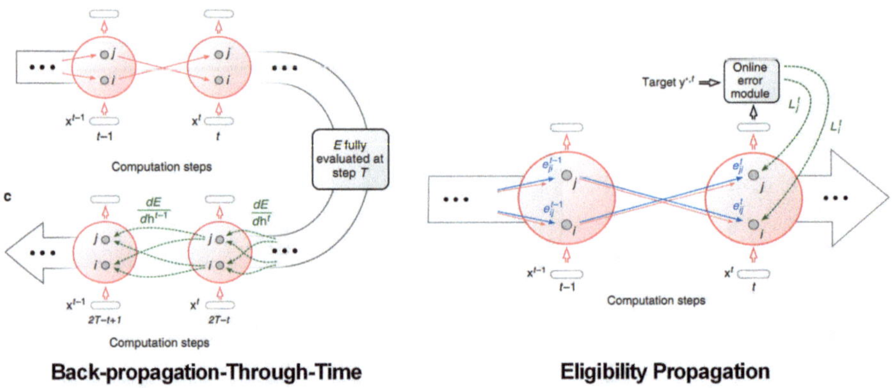

Fig. 2.37 Overview of E-prop (Source: Guillaume Bellec, "A solution to the learning dilemma for recurrent networks of spiking neurons")

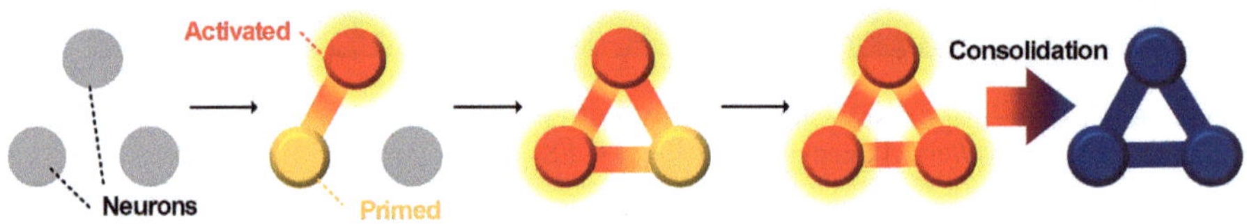

Fig. 2.38 Overview of Hebbian learning

However, it does not mean local learning is only restricted within unsupervised learning boundaries (Fig. 2.37).

Eligibility Propagation (E-prop)

The E-prop [150] focused on back-propagation through time that is usually used for RNN training. Since BPTT requires neurons and time to transmit the gradient backward, it is not claimed to be physiologically feasible. In this chapter, we present e-prop, a novel gradient-based learning technique that uses local and online weight update rules for recurrent neural networks, especially recurrent spike neural networks (RSNNs). To increase the computational effectiveness of BPTT, e-prop introduced an approximation technique that avoids neural state preservation. A mathematical approximation to BPTT is the local and online learning method for RSNN known as E-prop.

The fact that the accurate computation of qualification tracking relies on numerous hidden states of neurons is an important feature that sets the e-prop apart from previous qualification tracking-based algorithms. Due to the fact that neurons can have multiple hidden states, there are different types of neuron models accessible in e-prop. With e-prop, real numbers are propagated directly to neurons with random weights, simulating neuromodulator function in the brain. The training signal is a local variant of the random broadcast sort (Fig. 2.38).

Hebbian Learning

One of the earliest modern brain theorists, Donald Hebb, approached psychology from a biological point of view. He blended behavioral and neuropsychological psychology techniques. His goal was to better understand how neurons work in mental functions such as learning. Hebbian learning [151, 152], or more formally Hebbian theory, is a way to explain the rules of fluctuation of synaptic weights between two neurons.

Hebbian learning can be summarized as follows: "When cell A functions as a sufficient and persistent cause to excite cell B, the force of cell A's action on cell B changes and increases internally."

Dogs salivate just looking at food. Simply put, cell C promotes cell B in Hebb's description of this metabolic process. Keep practicing and your dog will start salivating when you press the bell after feeding. In the end, A cells affect B cells, so even if you press the bell without giving food, saliva begins to pass. It can be said that the structure of the brain or the connection structure of the neurons that make up the brain was modified in the dog, causing bells and drooling responses. Even if the original structure of the brain has nothing to do with it. Hebbian learning is the name given to this type of education.

The earliest and most basic type of learning rule is the Hebbian learning rule based on Hebb's learning theory. This rule is based on the discovery that when neurons on either side of a synapse are frequently stimulated simultaneously, the strength of the connection between two neurons increases. I have the following mathematical expression:

$$\boldsymbol{w}_{i,j}(t+1) = \boldsymbol{w}_{i,j}(t) + \eta \boldsymbol{y}_j(t)\boldsymbol{x}_i(t) \tag{2.1}$$

where $\boldsymbol{x}_i$ and $\boldsymbol{y}_j$ denote the outputs of neurons i and j and $\boldsymbol{w}_j$ denotes the synaptic connections between them. η is the learning rate. Also, $\boldsymbol{x}_i$ is another input to the combined $\boldsymbol{w}_{i,j}$. Learning occurs locally, as only the state of two adjacent neurons can affect synaptic changes.

Predictive Coding

According to the predictive coding paradigm [153], our brain mainly functions to predict the sensations that will be input in the future. The brain learns to predict future sensory information in advance and realizes the prediction by assuming that a baseball in the sky acts as a present sensory input. When you have access to these forecasts, you can prepare your activities. As a hypothesis itself, the predictive coding paradigm is biologically sound. This is due to the fact that the senses immediately enter the "correct response" data for learning.

When time really goes into the future, prediction is a sense of the future, so you can judge the error by comparing sensory input with the sense you had previously predicted. We can learn lessons from mistakes made in this way and improve our

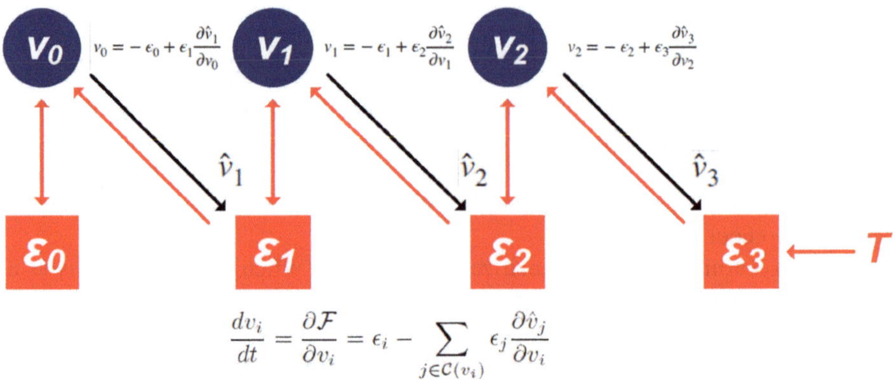

Fig. 2.39 Overview of predictive coding

forecasts. Through this process, our brain acquires the knowledge necessary for learning by itself. The ability of the predictive coding paradigm to do this is a key characteristic. In order for an artificial neural network to learn to spot mistakes and back-propagate through them, some kind of "correct answer" must be provided. However, there is no such thing as a "correct response" in real life. In the predictive coding paradigm, this kind of "correct response" is itself presented by the senses. Another name for this trait is self-supervised learning.

What we are going to do next is apply predictive coding to our neural network. A method for approximating back-propagation through predictive coding between adjacent neurons was proposed in James CR Whittington's 2017 paper An Approximation of the Error Back-propagation Algorithm in a Predictive Coding Network with Local Hebbian Synaptic Plasticity, and this method is demonstrated in a simple multi-layer perceptron. Extending this work, Beren Millidge's Predictive Coding Approximates Back-propagation along Arbitrary Computation Graphs (2020) shows that predictive coding learning rules between adjacent neurons can also successfully approximate back-propagation in addition to multi-layer perceptron, convolutional neural network, recurrent neural network, and long short-term memory, etc.

The chain rule does not fully apply to predictive coding. Only nearby neurons are used to generate the learning rule. For example, consider the connection between neurons v_1 and v_2. Based on the goal of predictive coding, neuron v_1 predicts what the value of v_2 will be in the future. If the value of v_2 is published a little later, an error ϵ_2 is raised. You can use this real number to adjust the connection weights for v_1 and v_2. Since v_1 and v_2 are really tightly coupled and only use information close to the link weight adjustment, it is not too difficult to explain biologically at this point. If these processes are performed in parallel at v_0 and v_1, v_1 and v_2, v_2 and output v_3, the entire process will eventually approximate the process of back-propagation (Fig. 2.39).

In this chapter, the actual implementation was constructed so that the prediction of peripheral neurons of the Gaussian generation model becomes a parameter. The specifics are beyond the scope of this chapter. In any case, there is no need to transmit real numbers generated in the output layer to the input layer via a global link. Because nearby neurons can learn at the same time and approximate back-propagation. In addition, the creation of these parallel learning algorithms is important for minimizing the time required for deep learning as well as biological viability. The more parts can be parallelized when we use multiple computers and it effectively accelerates the learning by shortening the process time.

Local Error Training

Local error based training [154, 155] is simply a learning method that teaches the weights of a specific layer using local real numbers generated using the layer's neural state variables. The input is the neural activation of the layer being trained, and the error is generated directly from the training labels using a classifier with fixed random weights and no hidden layers. As a result, training reduces local loss functions to more than a few global loss functions. Consequently, this method adjusts all parameters to achieve one of the fundamental principles of conventional back-propagation, integration loss minimization. Deep networks can transform features found at lower layers into more sophisticated features at higher levels, even after training with local mistakes. This increases the accuracy of the random local classifier at a deeper level, which can be seen below. One of the main benefits of deep networks is that the hierarchical arrangement of features is preserved by training with local values (Fig. 2.40).

As demonstrated in a previous study [155], the backward locking problem can be solved by learning the locally generated accidental hidden layers layer-by-layer.

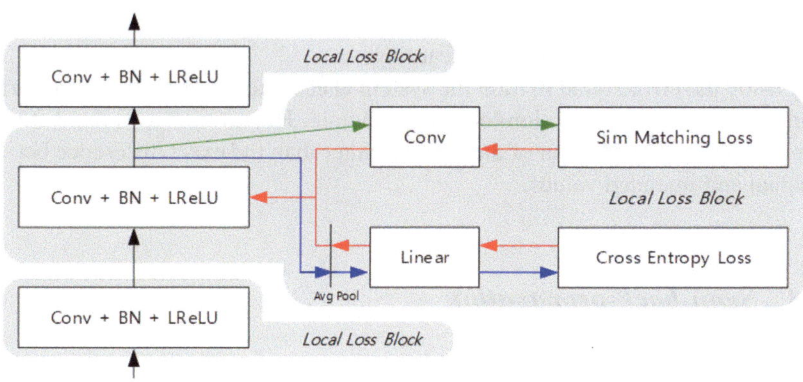

Fig. 2.40 Overview of local loss (Source: Arild Nøkland, "Training Neural Networks with Local Error Signals")

Local loss functions do not depend on globally generated real numbers, gradients do not go back to previous layers, and hidden layer weights can change as the algorithm progresses. The network behaves like a normal network trained using global back-propagation in the inference phase. Gradients and activations no longer need to be stored in memory when the weights of the hidden layer are modified. This reduces the amount of memory required to train the deep network. You can train all layers at the same time while training each layer one at a time due to local errors, further reducing memory footprint and speeding up training.

Error Correction Learning

When error correction learning [156] and supervised learning are combined, the output of the system is compared to a target output value, and then the error is used to guide training. You can directly adjust the tap weights using the error number using the same method as the shortest path back-propagation algorithm. If the system output is y and the expected system output is d, the error signal can be written as

$$\boldsymbol{e} = \boldsymbol{d} - \boldsymbol{y} \tag{2.2}$$

At each training iteration, the error correction learning algorithm tries to reduce this error signal.

Boltzmann Learning

The statistical concept of Boltzmann learning [157] comes from the study of thermodynamics. It is used for supervised learning and is similar to mistake correction learning. In addition to the overall output, the algorithm also considers the state of each individual neuron. In this respect, the Boltzmann learning rule is much slower than the error correction learning rule. A Boltzmann machine is a neural network that uses Boltzmann learning.

By using the error signal to train the system at each iteration, Boltzmann learning is similar to the error correction learning principle. However, it uses the difference in the probability distribution of the system rather than the exact difference between the actual and intended values.

2.8.3 Semi-back-propagation

Feedback Alignment

Feedback alignment [44] promises to solve the problem of adjusting the weight transfer feedback. In the back-propagation-based DNN training, the weights for the feedback path for layer n and the forward propagation path for layer $n + 1$ must be perfectly symmetric (weight symmetry condition). In practice, however, the publication asserts that such strict requirements are not necessary. If only one weak constraint is met, training can be performed without discerning back-propagation and performance differences, even if the weights in the feedback path are fixed to a fixed random number.

"Feedback alignment" means "constructing and supporting a chaotically occurring feedback signal." In other words, the weights of the forward propagation path automatically correct (align) the input feedback signal using the weights of the fixed random feedback path and use it for training. In other words, if you give the neural network arbitrary feedback, it automatically learns the feedback signal. No need to provide feedback with sophisticated values such as back-propagation.

The requirement that the weight matrix must match the weight matrix of the forward propagation path can be seen in the feedback alignment. However, even if we substitute the weight matrix of this feedback loop into matrix B set to a random value as shown in the figure on the right, the training remains the same. This is achievable because back-propagation's learning rules are heuristic and iterative from the start. You do not learn gradient descent all at once. The minimum error point can only be reached if multiple gradient descent algorithms are combined. In other words, when going down an incline, the direction is much more important than the speed at which you descend. Gradient descent is the method of choice for learning if the direction is correct.

According to feedback alignment research, this matrix B must obey one minor limitation. The numbers do not have to match exactly, but the directions must match for the training (updating weights) to be correct. The requirement is that the angle between B_e, a component of the feedback value used for feedback alignment, and We, a component of the feedback value used for the current back-propagation, must be less than 90°. After all, W and B should be as comparable as possible. Actually, this is a trivial condition. This is because W is constantly updated and aligned with B. This alignment procedure reduces angular mismatch. However, the error value e generated in the output layer is considered to be universally propagated in the feedback alignment. Real neural networks are severely limited by this assumption.

Direct Feedback Alignment

Even though both the back-propagation and the feedback alignment propagate errors from the last layer to the first layer in order, the direct feedback alignment ([71]) directly propagates errors from the last layer to other layers. If the number of neurons in the last layer and the ith layer is represented by N_L and N_i, respectively, the size of the feedback weight is determined as $N_i \times N_L$. Let ith layer's feedback weight in the direct feedback alignment be $\boldsymbol{D}_i^T$, and then the error is calculated as

$$\boldsymbol{e}_i = (\boldsymbol{D}_i^T \boldsymbol{e}_L) \odot f'(\boldsymbol{i}_{i+1}) \tag{2.3}$$

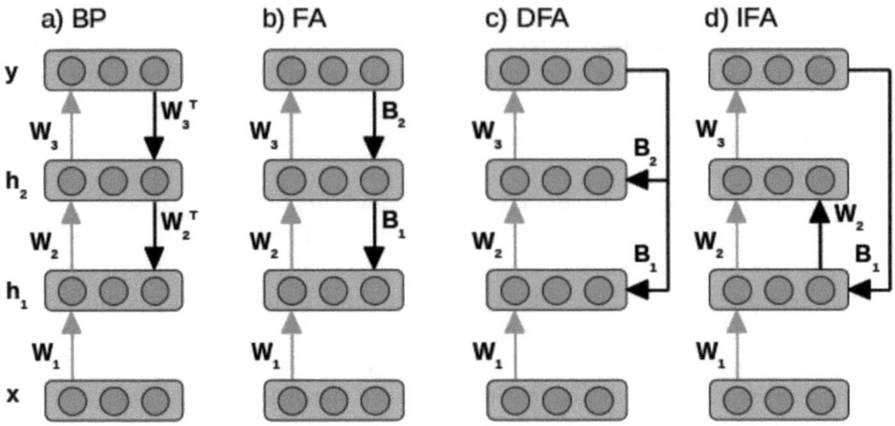

Fig. 2.41 Comparison of back-propagation, feedback alignment, and direct feedback alignment (Source: Arild Nøkland, "Direct Feedback Alignment Provides Learning in Deep Neural Networks")

One of the interesting characteristics is that there is no data dependency between different errors because the direct feedback alignment propagates the errors directly from the last layer. This characteristic gives the opportunity of parallel processing in the direct feedback alignment based error propagation operation. Compared with the back-propagation, direct feedback alignment showed similar training performance for the multi-layer perceptron structure. Moreover, direct feedback alignment requires fewer computations because N_L is generally smaller than the number of neurons in the intermediate layers (Fig. 2.41).

Decoupled Delayed Gradient

With decoupled parallel back-propagation using delay gradient, only the reverse pass parallelizes the pipeline. Forward passes are executed sequentially. By successively executing forward passes, an appropriate value can be utilized for the input–output value of the layer assigned to each compute node. However, in the pipelined approach, delayed gradients are used because the backward passes occur simultaneously. As the number of computer nodes increases, the effect of the delay gradient increases, limiting the extent to which parallelization can improve performance. Decoupled delayed gradient [69] divides DNN into multiple groups and performs back-propagation only within a group to loosen the backward locking problems. The gradient calculated by each group is transferred to the prior group at the next iteration. Since it adopts an inter-group pipeline during the back-propagation, multiple gradients can be computed at the same time, but each group utilizes the inference results obtained at a different time step. Thus, it should copy the weights of the multiple different time steps for parallel gradient computation (Fig. 2.42).

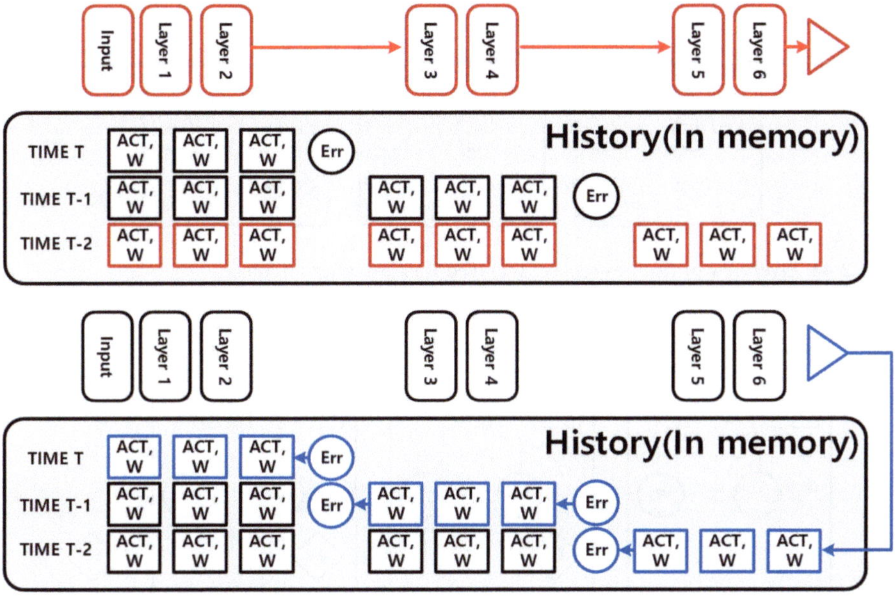

Fig. 2.42 Overview of decoupled delayed gradient (red: feed-forward, blue: error propagation)

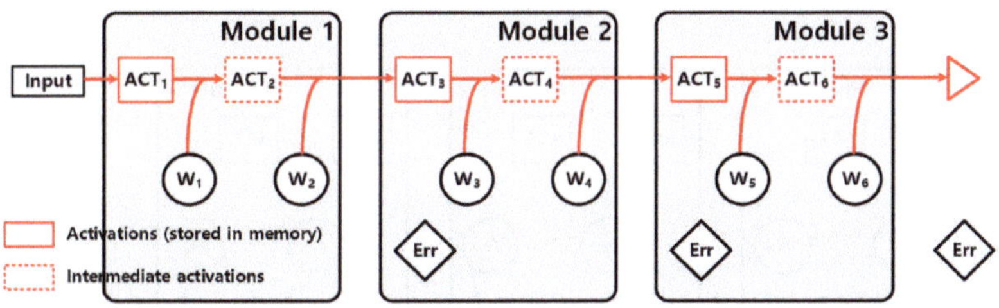

Fig. 2.43 Overview of features replay (1-1) feed-forward: inference of current inputs

Features Replay

Features replay [129] describes the difference between the exact gradient value and the delayed gradient value as a loss function in features replay and provides an algorithm that minimizes the loss function by utilizing the value of the forward pass delayed than the current value of the forward pass (Fig. 2.43, 2.44, 2.45, 2.46, 2.47, and 2.48).

Decoupled Neural Interface

Decoupled neural interface (DNI, [111]) proposes asynchronous management of data injection and gradient updates between neural network layers.

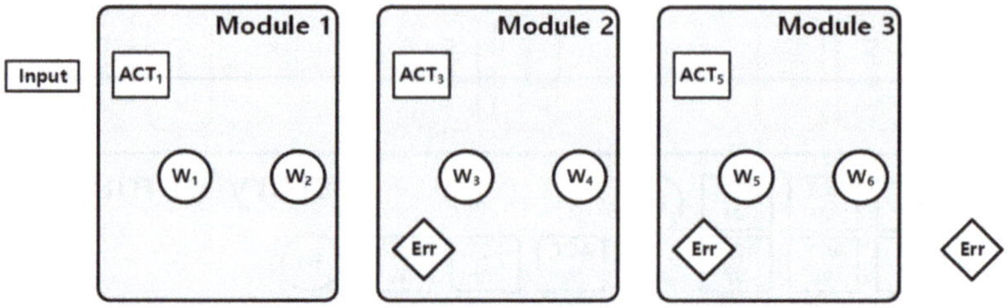

Fig. 2.44 Overview of features replay (1-2) feed-forward: after inference

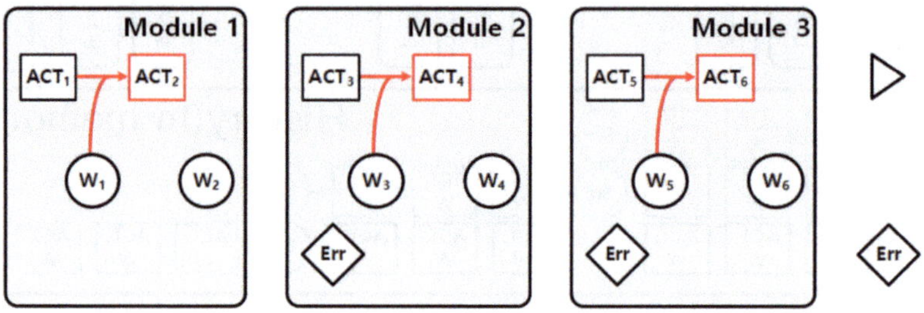

Fig. 2.45 Overview of features replay (2-1) backward path: replay for delayed inputs

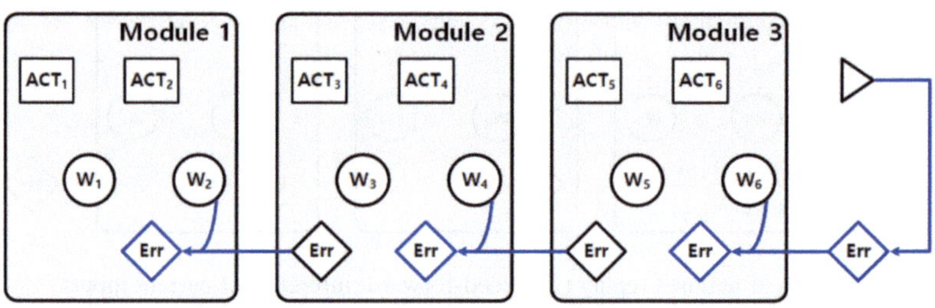

Fig. 2.46 Overview of features replay (2-2) backward path: module-to-module error propagation

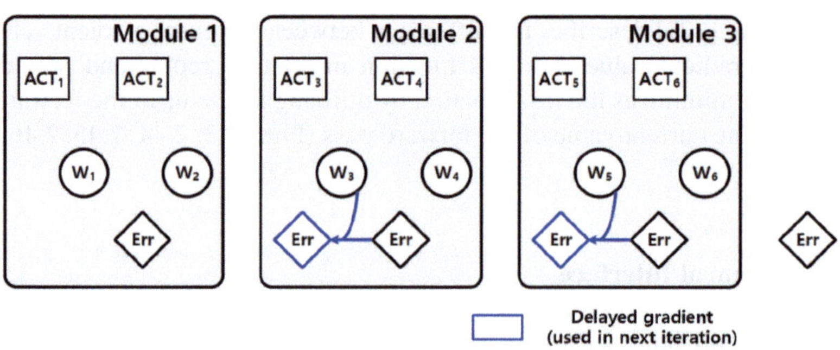

Fig. 2.47 Overview of features replay (2-3) backward path: error propagation inside the module

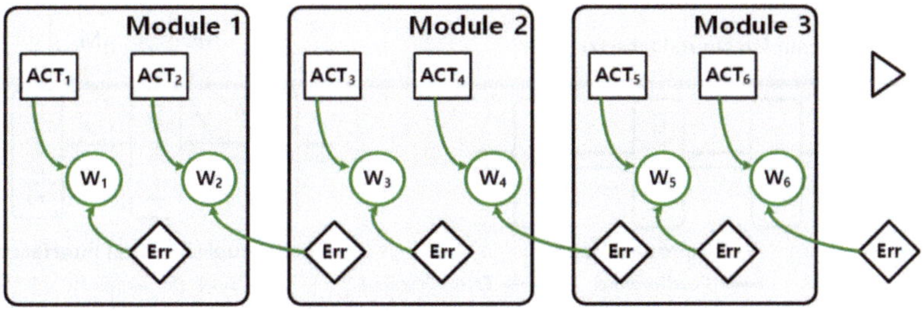

Fig. 2.48 Overview of features replay (3) weight-gradient-update

We can say that one layer of the neural network processes the data and the other layer is locked. In particular, when an inter-layer operation is transferred to another device, it is physically terminated. Error gradient back-propagation follows the same rules. Due to the nature of the chain law, if no gradient is passed from the upper layer, the lower layer will be in the standby state. Since synchronization between layers affects the learning performance of large-scale neural networks, we propose a methodology to asynchronously manage data flow between layers.

Instead of waiting for the data to be processed in one layer and then sent through another layer to compute and back-propagate the errors of the last layer, the approach here suggests updating the composite gradient as soon as the data are processed. The following mini-batch data are accepted after the layer's parameters have been updated with a false gradient. When the real gradient is sent from the previous layer, the false gradient learns to use this value to create a better fake gradient. In other words, there is a small neural network between the layers to generate artificial gradients. The asynchronous update procedure for these gradients is shown in Fig. 2.49.

Fully Decoupled Gradient

Another method, fully decoupled gradient [70], copies intermediate input activations instead of weights. Not only error propagation but also the feed-forward stage is divided into multiple groups and makes a big pipeline structure among them. Each group updates its weight with the delayed input activations and performs a feed-forward operation as soon as it completes the weight update. It can further decrease processing time thanks to the pipeline structure of both the feed-forward and error propagation stages (Figs. 2.50 and 2.51).

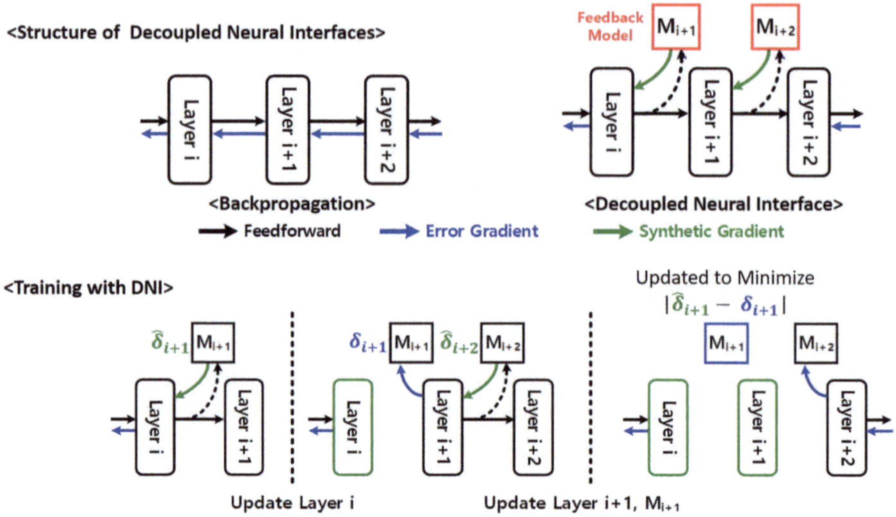

Fig. 2.49 Overview of decoupled neural interface

2.9 Relationship Between Cloud Server and Mobile Devices During the Learning

2.9.1 3 Types of Learning Based on Its Training Platforms

Server-Only Learning

This is a naive DNN training methodology. Although it can collect training datasets easily from user domains, it has critical privacy issues. Users need to transmit their private data to the cloud server to increase the dataset adopted in the cloud server. Nobody wants this kind of training methodology.

On-Device (Device-Only) Learning

On-device learning is the ideal purpose of AI research. Cloud server just gives a pre-trained model that has general knowledge based on a large dataset. However, the real-world user environment is biased, and the pre-trained model with high generality sometimes shows poor performance. To fit users' demands, on-device learning is a great solution for adaptation. Since the data collected on a mobile device do not need to transmit to a cloud server, it is free from privacy issues. However, complete on-device learning is highly challenging. It is not that easy to be implemented in the near ages.

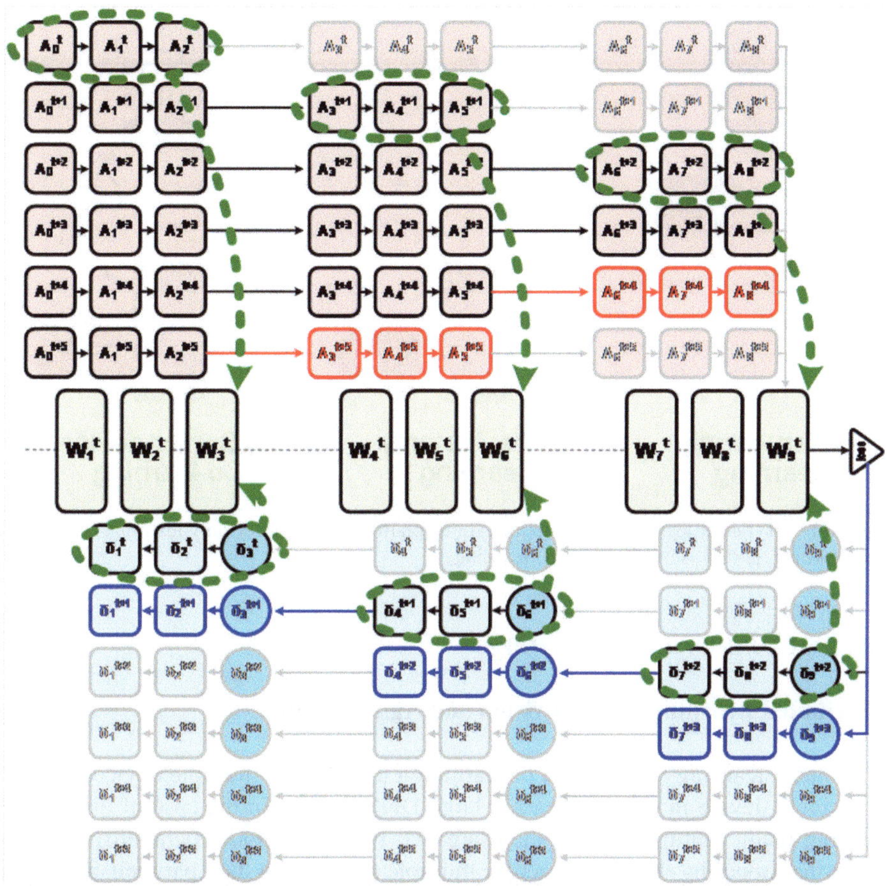

Fig. 2.50 Overview of fully decoupled gradient

Server–Device Co-learning

Due to the difficulty of complete on-device training, server–device co-learning can be an alternative. Encrypted data transmission can be useful to get AI applications with great performance. The data obtained from the cloud server can be either trained models or labels. Moreover, collaborative learning of the server and each user's device can enlarge the DNN training scale to make AI with global, general knowledge. The remaining parts will explain the various types of collaborative learning.

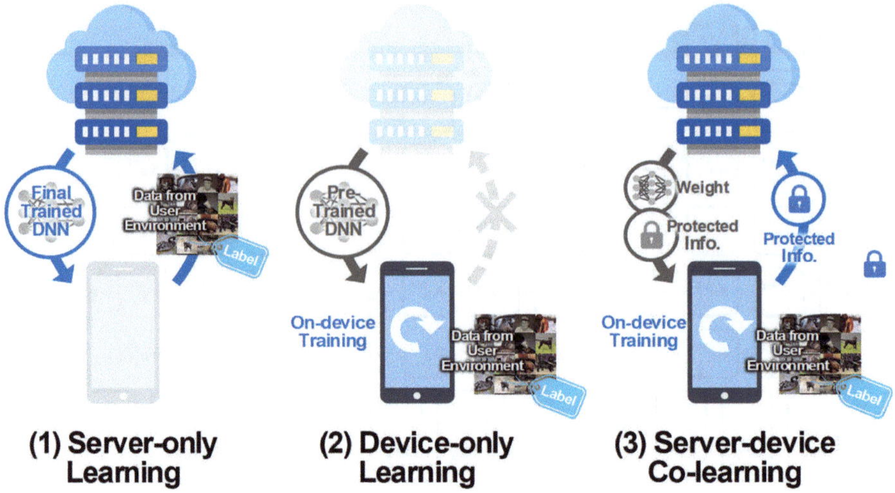

Fig. 2.51 3 Types of learning based on its training platforms: (1) server-only learning, (2) device-only learning, and (3) server–device co-learning

2.9.2 Collaborative Learning

Distributed Learning: Task Division and Aggregation

Distributed learning is a method of learning a model with multiple devices instead of a single device in order to learn a large DNN model. In the case of distributed learning using multiple GPUs, there are two methods: model parallelism, which trains the model by dividing it across multiple GPUs, and data parallelism, which learns by dividing the data to be trained. In the case of distributed learning between devices and devices, rather than the method of learning by dividing the model, there are many forms of data parallelism that share the results learned from each data based on the numerous data they collect in each environment.

This data parallelism not only protects privacy but also considers the learning results together and creates a global model because learning is carried out on each device without the need to transmit the data collected by each user device. The generality of the model can also be greatly increased. Data parallelism has the advantage of shortening the learning time because multiple devices perform operations separately compared to the method of learning by moving all data to a single device. However, the disadvantage is that it requires synchronization of the divided workloads. It needs to synthesize the learning results each time when the weight parameter is updated. Since the training data are divided, the direction in which the weight (gradient) parameter is updated varies according to the training data received. Therefore, synchronization, which is the process of collecting all gradients created by multiple devices, averaging them, and then dividing them again, is essential. After synchronization, the next training data are distributed based on the same model (parameter), and training starts.

This form of distributed learning can be server–device, and joint training between device–device is also possible. However, in order to reduce the burden of synchronization, a method that can minimize the cost of gradient communication is absolutely necessary.

Low-Cost Gradient Transmission Methodologies

1. **Gradient Bit-Precision Reduction**: Gradient compression is important to reduce the communication cost induced by synchronization. Seide et al. [60], Alistarh et al. [61], Wen et al. [62] suggested ultra-low-bit gradient quantization, but their main purpose was to reduce the gradient communication among the multiple-edge devices during the distributed learning. Since the primal weight still needs full-precision (16-bit), the quantized gradient should be converted into the original bit-width during the weight update operation. In addition, a large portion of memory access is still required due to the data movement of the primal weight and momentum from the memory.
2. **Sparse Gradient Selection**: Gradient compression can be realized by not only quantization but also gradient sparsification. Lin et al. [63], Strom et al. [64] accumulated weight gradients every iteration and updated only when the accumulated gradient became larger than the pre-determined threshold. These methodologies require a large internal buffer to store all weight gradients, which is not practical in training on micro-AI systems. Another sparsification [65] updated the weight only when it had a large gradient or the target confidence was low [66]. Large-gradient-only or low-confidence-only weight update showed great compression ratio, but they can cause a slow training curve. Furthermore, they are not easy to be applied to other complicated DNN applications such as object detection and image-to-image generation.

Federated Learning

Sometimes federated learning is included in distributed learning category, but it shows distinct characteristics compared with the conventional distributed learning concepts. Without explicitly sharing data samples, federated learning attempts to train deep neural networks on various local datasets stored on local nodes. To create a global model shared by all nodes, a common method involves training a local model on local data samples and transferring "parameters" (deep neural network weights and biases) between these local nodes at specific frequencies.

As federated learning is designed to learn on heterogeneous datasets and distributed learning is designed to parallelize compute capacity, the main difference between the two approaches lies in their assumptions about the nature of the local

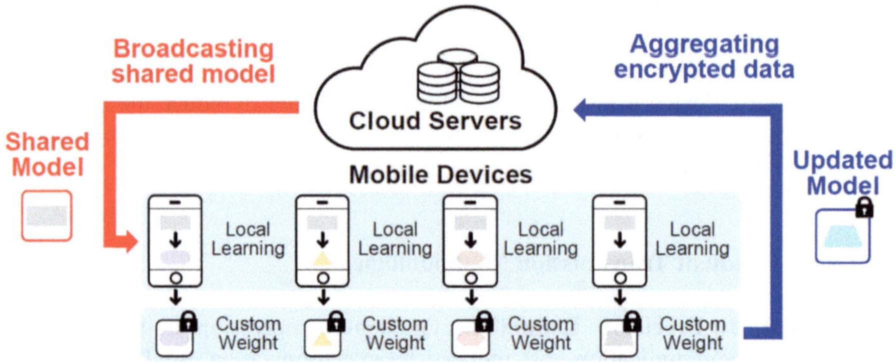

Fig. 2.52 Overview of federated learning

dataset. A basic assumption that often occurs in distributed learning is that local datasets are independent, uniformly distributed, and approximately the same size. Distributed learning also tries to train a single model on many servers. Federated learning is not an intended use of these theories. Instead, datasets can vary many times in size and are usually diverse. Additionally, clients using federated learning experience more failures or dropouts more frequently due to their reliance on weak communication tools and battery-powered devices, so they may be trusted or not (Fig. 2.52).

Assuming that there is only one iteration of the learning process in a federation round, the learning process can be summarized as follows:

1. **Initialize**: Choose a machine learning model (such as linear regression, neural network, or boosting) to train and initialize on a local node based on server input. It then waits for the node to become active and the main server to start processing compute tasks.
2. **Client Selection**: A subset of local nodes is selected to begin training on local data. The selected node purchases the current statistical model, and the other nodes wait for the federation to resume.
3. **Configuration**: Selected nodes are instructed by a central server to perform model training on local data in a specific way (e.g., updating some mini-batch of gradient descent).
4. **Reporting**: Each node selected for aggregation sends its local model to the server. The central server compiles the received model and propagates the model changes to the nodes. It also manages issues related to unconnected nodes or missing outdated models. Return to the client selection phase and the next round of federation begins.
5. **Termination**: The central server collects updates and completes the global model when pre-determined shutdown requirements are met (for example, when the maximum number of iterations is complete or the model accuracy is higher than a threshold).

The previously thought method assumes synchronous model updates. New methods for overcoming asynchrony throughout the training process or during training with dynamically changing models have recently been developed in federated learning. An asynchronous approach uses the properties of a neural network to exchange model updates as soon as computations in a particular layer are available, as opposed to a synchronous approach in which local models are exchanged after computations for all layers of the neural network are complete. These methods, also known as split learning, can be used for both training and inference processes, whether federated learning is distributed or centralized.

Gossip Training

An entirely new distributed training technique called "gossip training" uses a randomized gossip algorithm to reduce training delays. Gossip averaging [158] is a pioneering study of random gossip algorithms that allow peer-to-peer information sharing to converge rapidly toward consensus across nodes. Because there are no prerequisites for centralized nodes or variables, the gossip distribution method benefits from total asynchrony and full decentralization.

Gossip SGD [159] and GossipGraD [160] are the latest developments in gossip education. A distributed and asynchronous training method for DNN models called gossip SGD has been proposed. Gossip SGD oversees a separate collection of nodes, with each node iteratively running the DNN model in two steps (gradient update and mixed update). The hosted DNN model is updated locally by each node in the gradient update phase, where information is shared with additional randomly selected nodes in the blended update phase. When all DNNs reach a consensus, the process repeats.

Gossip SGD is an effort to update SGD using the gossip algorithm, but also with significant performance loss. Deep learning methods can be scaled up on large systems with the help of the SGD algorithm known as GossipGraD, which is based on the gossip communication protocol. According to GossipGrad, diffusion is thought to lower the overall cost of communication, which also indirectly allows computing nodes to trade updates (gradients) every few steps. To promote direct gradient spread and prevent overfitting during the feed-forward phase of SGD, we additionally consider the rotation of the communication partner (Figs. 2.53 and 2.54).

Temporal Knowledge Distillation

Temporal knowledge distillation suggested by [42] is a concept of a combination of knowledge distillation and online learning concepts. Temporal knowledge distillation utilizes a large teacher network as a label generator and trains a small student network by using the knowledge distillation concept. Farhadi et al. [42] demonstrates online DNN training with object detection applications. Although the

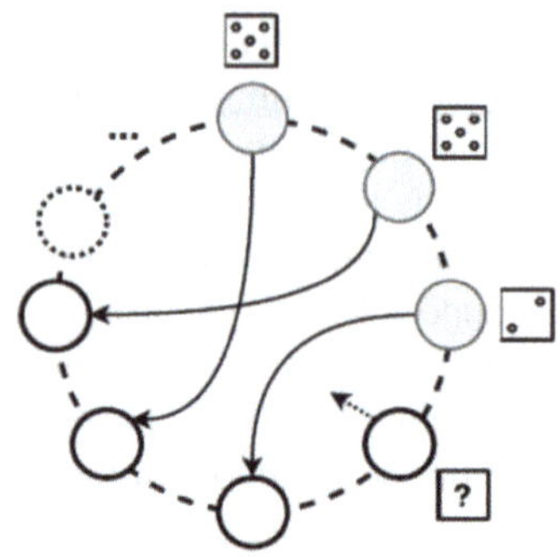

Fig. 2.53 Communication with randomly selected partner in gossip manner (Source: Z. Zhou et al., "Edge Intelligence: Paving the Last Mile of Artificial Intelligence With Edge Computing")

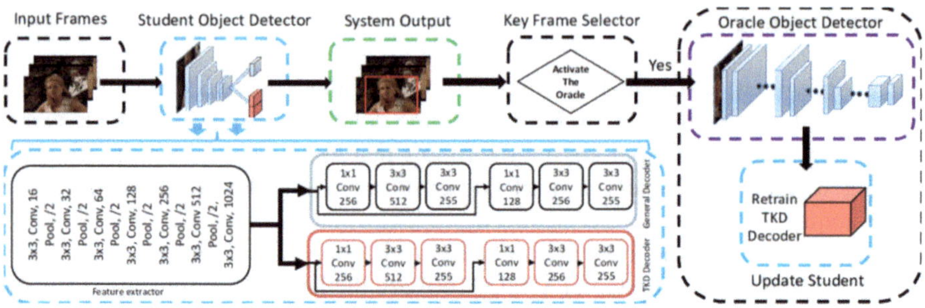

Fig. 2.54 Overview of temporal knowledge distillation based object detection (Source: M. Farhadi et al., "TKD: Temporal Knowledge Distillation for Active Perception")

small student network is more useful to be adopted in mobile devices, it shows poor performance compared with large networks. Farhadi et al. [42] adopts YOLO-v3 as a teacher network and offers soft labels to tinyyolo-v3 network. Since the teacher network is sparsely used, the cost of large network inference can be reduced but still shows good performance by tuning the student network.

This temporal knowledge distillation is one of the great solutions to offer labels to mobile/edge devices but is still challenging because it needs to accelerate both the teacher network and student network. Although the teacher network is rarely used, the mobile device should have enough memory space to store the parameters of a large teacher network. Although both HNPU-V1 and HNPU-V2 adopt the on-device training concept, it can be enlarged to server–device co-training to make efficient but accurate AI applications. When we assume the acceleration of the teacher network is the role of the cloud server, the mobile device can remove the cost of teacher network acceleration. The problem is, it needs to transmit data collected by a user-specific environment. This method finally induces critical privacy issues. In spite of this problem, temporal knowledge distillation based server cloud co-training is still attractive to minimize the burden of large neural networks in mobile devices. Instead, it should be co-researched with privacy-preserving techniques such as data encryption.

Chapter 3
New Algorithm 1: Binary Direct Feedback Alignment for Fully Connected Layer

There were many algorithms to substitute the back-propagation (BP) in the deep neural network (DNN) training. However, they could not become popular because their training accuracy and the computational efficiency were worse than BP. One of them was direct feedback alignment (DFA), but it showed low training performance especially for the convolutional neural network (CNN). In this chapter, we overcome the limitation of the DFA algorithm by combining with the conventional BP during the CNN training. To improve the training stability, we also suggest the feedback weight initialization method by analyzing the patterns of the fixed random matrices in the DFA. Finally, we propose the new training algorithm, binary direct feedback alignment (BDFA), to minimize the computational cost while maintaining the training accuracy compared with the DFA. In our experiments, we use the CIFAR-10 and CIFAR-100 datasets to simulate the CNN learning from the scratch and apply the BDFA to the online learning based object tracking application to examine the training in the small dataset environment. Our proposed algorithms show better performance than conventional BP in both two different training tasks especially when the dataset is small.

This chapter is based on paper "Efficient Convolutional Neural Network Training with Direct Feedback Alignment" (ICCVW or arXiv:1901.01986 [72]).

3.1 Introduction

The deep learning becomes the core of the machine learning, and it has been utilized for many applications such as machine translation [85], speech recognition [86], and object classification [87]. Training the deep neural network (DNN) is an important portion of the deep learning because we need different pre-trained models to cover the various deep learning tasks. One well-known method of DNN training is the algorithm called back-propagation (BP) [40]. The BP is the gradient descent based

© The Author(s), under exclusive license to Springer Nature Switzerland AG 2023
D. Han, H.-J. Yoo, *On-Chip Training NPU - Algorithm, Architecture and SoC Design*,
https://doi.org/10.1007/978-3-031-34237-0_3

training method that follows the steepest gradient to find the optimum weights. Therefore, BP based training can be applied to any DNN configurations if the network consists of any differentiable operations. For instance, not only multilayer perceptron (MLP) but also both convolutional neural network (CNN, [88]) and recurrent neural network (RNN, [89]) can be trained by using the BP.

Even though the BP shows the outstanding performance in DNN training, BP based training suffers from the overfitting problem. Since the BP easily sinks into a local minimum, we need a large scale of the dataset to avoid the overfitting. In addition, we use various data augmentation techniques such as flipping and cropping. If the DNN training is done in the limited resources and dataset, BP based DNN training is too slow to be converged and shows low accuracy. One example is MDNet [41] that introduces an object tracking algorithm with the online learning concept. Real-time implementation is an important issue in object tracking, so it has the limitation to utilize various data augmentation and large dataset for online learning.

To break the BP based DNN training paradigm, a lot of new training methods have been developed. Salimans et al. [90] proposed the evolution strategy that searches various weights without the gradient descent. It has a chance to be computed in parallel, but it needs a lot of seeds to find the global optimal solution. Moreover, evolution strategy has a slow convergence problem that can be an obstacle for fast online learning application. Another algorithm, feedback alignment (FA, [91]), was proposed based on the gradient descent methodology, but without following the steepest gradient. The FA pre-defines a feedback weight before starting the training, and the weight is determined by the random values.[1] It is also known to converge slowly and shows worse performance compared with the previous BP.

Direct feedback alignment (DFA, [71]) was developed by getting the idea from the FA. Although the FA propagates errors from the last layer back to the first layer step by step, the DFA propagates errors from the last layer directly to each layer. This approach gives the opportunity of the parallel processing. Since the errors of every layer are generated independently in DFA, we can immediately calculate each layer's gradient if the DNN inference is finished. Moreover, the number of required computation is reduced in the DFA. This is because the number of neurons in the last layer is usually fewer than the prior layer, so the size of the feedback weight becomes smaller than the BP. In spite of these advantages, the DFA suffers from accuracy degradation problem. The accuracy degradation problem becomes more serious when the DFA is applied to the CNN case. It is known that the DNN is not learnable with the DFA if the DNN becomes much deeper than AlexNet [93].

In this chapter, we explain the new DFA algorithm to improve the training accuracy in CNN and suggest a feedback weight initialization method for the fast convergence. Moreover, we propose the binary direct feedback alignment (BDFA) to maximize the computational efficiency. We verified the training performance in

[1] The feedback weight was notated as a fixed random matrix in the original paper.

the VGG-16 [94] without any data augmentation, and DFA shows higher accuracy compared with the BP. And then, the training with small dataset was proved through the online learning based object tracking application. Our proposed algorithm shows better performance than the conventional BP approach in both the learning from the scratch and the object tracking tasks. The remaining part of the paper is organized as follows. The mathematical notation of the BP, FA, and DFA will be introduced in Sect. 3.2. Then, the details of the proposed algorithms will be explained in Sect. 3.3. The experiment will be followed in Sect. 3.4, and the paper will be concluded in Sect. 3.5.

3.2 Preliminaries

3.2.1 Back-propagation

Back-propagation is a general algorithm for the DNN training, suggested by Rumelhart et al. [40]. Let $\boldsymbol{o}_i$ be the ith layer's feature map, when $\boldsymbol{W}_{i+1,i}$ be the weight and bias between the ith layer and the $i+1$th layer. If the activation function is represented as $f(\cdot)$, and L is the total number of layers, the feature map of the $i+1$th layer can be calculated as

$$\boldsymbol{i}_{i+1} = \boldsymbol{W}_{i+1,i}\, \boldsymbol{o}_i, \quad \boldsymbol{o}_{i+1} = f(\boldsymbol{i}_{i+1}), \quad i \in \{0, 1, \ldots, L-1\} \tag{3.1}$$

Various activation functions such as sigmoid, tanh, and ReLU [95] can be one of candidates of the $f(\cdot)$. Some DNNs such as generative adversarial networks [96] use more advanced activation functions such as Leaky ReLU and PReLU [97]. Once the inference is over, the inference result is compared with the pre-defined labels, and the error map $\boldsymbol{e}_L$ is calculated by a loss function such as cross-entropy. The error is propagated gradually from the last layer to the first layer, and the ith layer's error map can be calculated as

$$\boldsymbol{e}_i = (\boldsymbol{W}^T_{i+1,i}\, \boldsymbol{e}_{i+1}) \odot f'(\boldsymbol{i}_{i+1}), \quad i \in \{1, 2, \ldots, L-1\} \tag{3.2}$$

where $\odot$ is an element-wise multiplication operator and $f'()$ is the derivative of the non-linear function. We need the transposed weight matrix, $\boldsymbol{W}^T_{i+1,i}$, to propagate the errors in MLP case. After the BP, the gradient of each layer is computed by using both the ith layer's feature map and $i+1$th layer's error map. The ith layer's gradient, $\boldsymbol{G}_i$, is calculated as following:

$$\boldsymbol{G}_{i,b} = \boldsymbol{e}_{i+1}\, \boldsymbol{o}_i^T, \quad i \in \{0, 1, \ldots, L-1\}, \quad b \in \{0, 1, \ldots, B\} \tag{3.3}$$

$$\boldsymbol{G}_i = \frac{1}{B} \sum_b \boldsymbol{G}_{i,b}, \quad \boldsymbol{W}'_{i+1,i} = \boldsymbol{W}_{i+1,i} - \eta\, \boldsymbol{G}_i \tag{3.4}$$

If we use the mini-batch gradient descent with the batch size B, total B gradients, $\boldsymbol{G}_{i,b}$, are averaged and get the $\boldsymbol{G}_i$. $\boldsymbol{G}_i$ is used to update the $\boldsymbol{W}_{i+1,i}$ by multiplying learning rate, η. In the CNN case, the matrix multiplication operations in every step described in MLP is substituted with the convolution operations. One more different thing is that it uses 180° flipped kernel instead of the transposed weight used in the BP. Other operations are as same as in the MLP case.

3.2.2 Feedback Alignment

Feedback alignment (FA) that was introduced by Lillicrap et al. [44] substituted the transposed weight with the fixed random matrix in error propagation operation. In other words, they pre-defined the feedback weight that has the same size as the transposed weight, but the values of the feedback weight are determined randomly. Although the weights used in the training are updated for every iteration, the pre-defined random matrix is maintained until the training is finished. Let $\boldsymbol{R}_{i+1,i}$ be the feedback weight of the ith layer, the propagated error in the FA is calculated as

$$\boldsymbol{e}_i = (\boldsymbol{R}_{i+1,i}^T \, \boldsymbol{e}_{i+1}) \odot f'(\boldsymbol{i}_{i+1}) \tag{3.5}$$

and the other procedures such as gradient generation and weight updating are same as BP does.

3.2.3 Direct Feedback Alignment (DFA)

Even though both the BP and the FA propagate errors from the last layer to the first layer in order, the DFA [71] directly propagates errors from the last layer to other layers. If the number of neurons in the last layer and the ith layer is represented by N_L and N_i, respectively, the size of the feedback weight is determined as $N_i \times N_L$. Let ith layer's feedback weight in the DFA be $\boldsymbol{D}_i^T$; then the error is calculated as

$$\boldsymbol{e}_i = (\boldsymbol{D}_i^T \, \boldsymbol{e}_L) \odot f'(\boldsymbol{i}_{i+1}) \tag{3.6}$$

One of the interesting characteristics is that there is no data dependency between different errors because the DFA propagates the errors directly from the last layer. This characteristic gives the opportunity of parallel processing in the DFA based error propagation operation.

Compared with the BP, DFA showed similar training performance for the multi-layer perceptron (MLP) structure. Moreover, DFA requires fewer computations because N_L is generally smaller than the number of neurons in the intermediate layers. However, the DFA dramatically degrades the accuracy when it is applied in

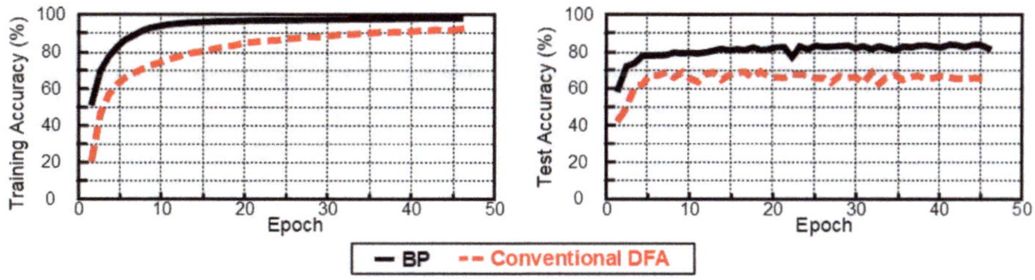

Fig. 3.1 BP (black) vs. conventional DFA (read) for CNN training (tested in CIFAR-10)

the CNN training as shown in Fig. 3.1. Furthermore, the elements of the intermediate feature map are all connected with the last layer neurons, so it requires much more computations compared with BP based CNN training. In summary, the DFA's computational efficiency can be induced in the MLP training, but not in the CNN.

3.3 Our Approach

As mentioned in Sect. 3.2, the DFA is efficient for MLP training because it can be computed in parallel without accuracy degradation. However, this advantage is diminished when the DFA is applied in the CNN. To solve this problem, we propose the new training method to make the DFA applied to the CNN. In addition, the initialization method of the feedback weight is suggested for the fast and stable learning curve. At last, we propose the advanced training algorithm, binary direct feedback alignment (BDFA), which shows high computing efficiency and robust training performance in various conditions.

3.3.1 CDFA: CNN Training by Combining both BP and DFA

Generally, CNN consists of the convolutional layers and fully connected (FC) layers. The two different kinds of layers have different roles. For example in object classification, convolutional layers are considered as the feature extractor by using the characteristics of the convolution computation. In contrast, FC layers receive the result of the feature extractor and judge what the object is. However, in the initial iterations of the BP based CNN training, training the convolutional layers can be ambiguous because the FC layers cannot be considered as a good object classifier. During some iterations of BP, the weight of the FC layers will be changed, and the convolutional layers should be adaptive to the changed FC layers. In other words, convolutional layers can be confused if the propagated error's domain is changed for every iteration.

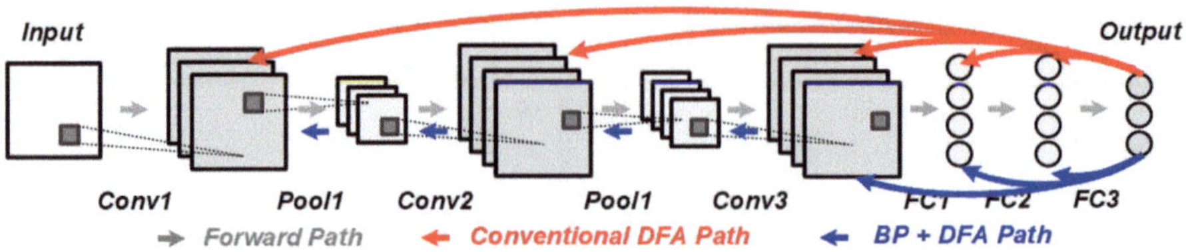

Fig. 3.2 Conventional DFA in CNN and proposed DFA based error propagation

If the error is propagated from the last FC layer to convolutional layer through the constant domain shifting, the training of the convolutional layer can be more stable than conventional BP. From this motivation, we use DFA instead of BP in the FC layers. As shown in Fig. 3.2, the network maintains the BP in the convolutional layers but adopts the DFA for the FC layers. Since there is no data dependency among the FC layers, the DFA can directly propagate the error to the FC1. Since the DFA uses fixed feedback weight for error propagation, convolutional layers do not have to be adaptive to various errors that are derived from the different domains. In this method, the error propagation in the convolutional layers can be started even though the errors are not propagated for the remained FC layers. Therefore, the error propagation of both the convolutional layers and the FC layers can be computed in parallel right after DFA is done for the first FC layer.

Nokland [71] shows that the randomly initialized feedback weight can be used for DFA based DNN training. However, both the FA and the DFA are sensitive to the initialization method of the feedback weight because it affects the training accuracy and the convergence speed significantly. As shown in [92], it is observed that the batch normalization (BN) seems to make the FA become not sensitive to the initialization method, but it is still a problem because of the slow convergence. To make the DFA robust to the initialization method, we fixed the feedback weight as the multiplication of the transposed weights in multiple layers. To sum up, the feedback weight of the ith layer, $\boldsymbol{D}_i$, can be calculated as

$$\boldsymbol{D}_i = \boldsymbol{W}_{L,L-1} \dots \boldsymbol{W}_{i+2,i+1} \boldsymbol{W}_{i+1,i}, \tag{3.7}$$

and finally, the error propagation operation can be summarized as follows:

$$\boldsymbol{e}_i = (\boldsymbol{D}_i^T \boldsymbol{e}_L) \odot f'(\boldsymbol{i}_{i+1}) = (\boldsymbol{W}_{i+1,i}^T \dots \boldsymbol{W}_{L-1,L-2}^T \boldsymbol{W}_{L,L-1}^T \boldsymbol{e}_L) \odot f'(\boldsymbol{i}_{i+1}) \tag{3.8}$$

The suggested initialization method is suitable for other various functions such as sigmoid, tanh, and ReLU. Moreover, other normalization or optimization methods such as BN and dropout [98] are also applicable with the proposed initialization with Eq. 3.7.

3.3.2 *Binary Direct Feedback Alignment (BDFA)*

The DFA needs the feedback weight addition to the forward weight matrix, and it occupies a larger memory to store both two matrices. Moreover, the DFA has the chance to be computed in parallel, but it requires much larger memory bandwidth. Since the throughput in the FC computing is vulnerable to the memory bandwidth, loading the additional feedback weight degrades the throughput compared with the unlimited bandwidth case.

To solve the throughput bottleneck problem caused limited bandwidth, we propose the binary direct feedback alignment (BDFA) algorithm. BDFA uses the binarized feedback weight, $\boldsymbol{B}_i$, whose values are determined as either $+1$ or -1. In other words, the $\boldsymbol{B}_i$ can be stored as a single bit to represent only the sign value of the feedback weight's element. As a result, required memory to store the $\boldsymbol{B}_i$ is reduced by 96.9% compared with the 32-bit floating point representation that now becomes the general numeric representation in CPU or GPU. As we determined in the DFA, BDFA's feedback weight, $\boldsymbol{B}_i$, can be similarly defined by a modification of Eq. 3.7. $\boldsymbol{B}_i$ is determined as

$$\boldsymbol{B}_i = sign(\boldsymbol{W}_{L,L-1} \dots \boldsymbol{W}_{i+2,i+1}\boldsymbol{W}_{i+1,i}), \quad \boldsymbol{e}_i = (\boldsymbol{B}_i^T \boldsymbol{e}_L) \odot f'(\boldsymbol{i}_{i+1}) \tag{3.9}$$

when the $sign(\cdot)$ is the function that indicates the sign value of each element. The difference between Eqs. 3.7 and 3.9 is only whether the $sign(\cdot)$ is applied or not. By applying Eq. 3.9, BDFA shows faster and stable training convergence compared with the random initialization case. The effect of the binarization and initialization will be discussed in Sect. 3.4.

3.4 Experiments

In this section, we compared the training accuracy of the conventional BP and suggested training algorithm. We measured the relative accuracy by training CNN from the scratch in CIFAR-10 and CIFAR-100 datasets [99]. We used the network configuration as described in Fig. 3.3. The base network follows the VGG-16 [94] configuration but has one additional FC layer. To sum up, it consists of 13

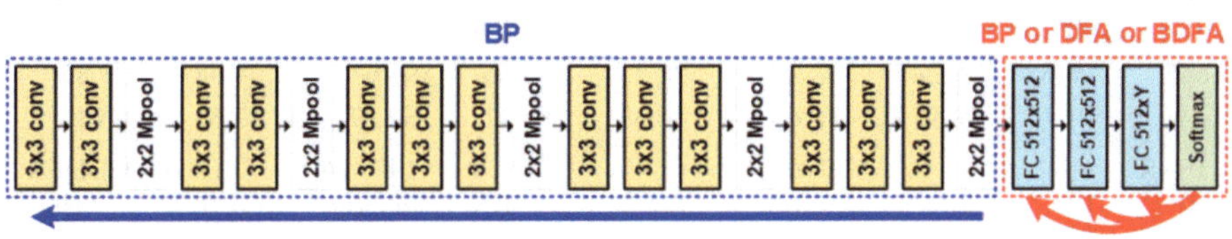

Fig. 3.3 Overall network configuration for training

convolutional layers with BN and ReLU activation functions, followed by three FC layers without BN. The number of neurons in the last FC layer is determined by the number of classification categories in each different dataset. In the BP based approach, both the convolutional layers and the FC layers are trained by using BP. In contrast, the training method of the last three FC layers is substituted with the DFA or the BDFA to measure the performance of the proposed training algorithm. The simulation was based on mini-batch gradient descent with the batch size 100 and uses momentum [100] for the optimization method. The parameters of the network are initialized as introduced by He et al. [101], and the learning rate decay and the weight decay method is adopted. Other hyper parameters are not changed for fair comparison.

3.4.1 CNN Training from the Scratch

The CNN training with the DFA and BDFA are renamed as CDFA and CBDFA, respectively, for the simple explanation. Table 3.1 shows Top5 and Top1 test accuracy after the CNN training is done in the two different CIFAR datasets. In this simulation, there is no data augmentation to make an environment that has a limited dataset. This condition can examine whether the algorithm is robust to the training in the small dataset. To sum up, only 50,000 images in the CIFAR-10 and CIFAR-100 are only used for DNN training, and the other 10,000 images are tested to evaluate the test accuracy. As a result, both the CDFA and the CBDFA show higher test accuracy compared with the conventional BP even though the feedback weight is randomly initialized. In CIFAR-10, the CDFA and CBDFA are 7.5% and 8.3% higher in Top1 test accuracy than the BP, respectively. The accuracy improvement by the CDFA and the CBDFA seems much more remarkable in CIFAR-100. As shown in Fig. 3.4, the training curve of the CDFA and CBDFA is much slower, but they achieve 21.3% and 19.6% better performance, respectively, compared with the BP.

However, the feedback weight with the random initialization has critical problems for training. One of the problems is the slow training curve described in

Table 3.1 CNN training result in CIFAR-10 and CIFAR-100 (small learning rate)

CIFAR 10	BP	BP w/ BN	CDFA random	CDFA Eq. 3.7	CDFA w/ BN	CBDFA random	CBDFA Eq. 3.9	CBDFA w/ BN
Top5	98.63	98.24	98.42	98.55	98.56	98.63	98.88	98.83
Top1	81.11	76.91	88.68	86.36	87.41	89.39	87.65	86.46
CIFAR 100	BP	BP w/ BN	CDFA random	CDFA Eq. 3.7	CDFA w/ BN	CBDFA random	CBDFA Eq. 3.9	CBDFA w/ BN
Top5	67.80	63.91	77.05	72.82	77.55	75.07	71.92	76.85
Top1	40.29	37.80	61.42	48.24	55.11	59.92	47.48	54.47

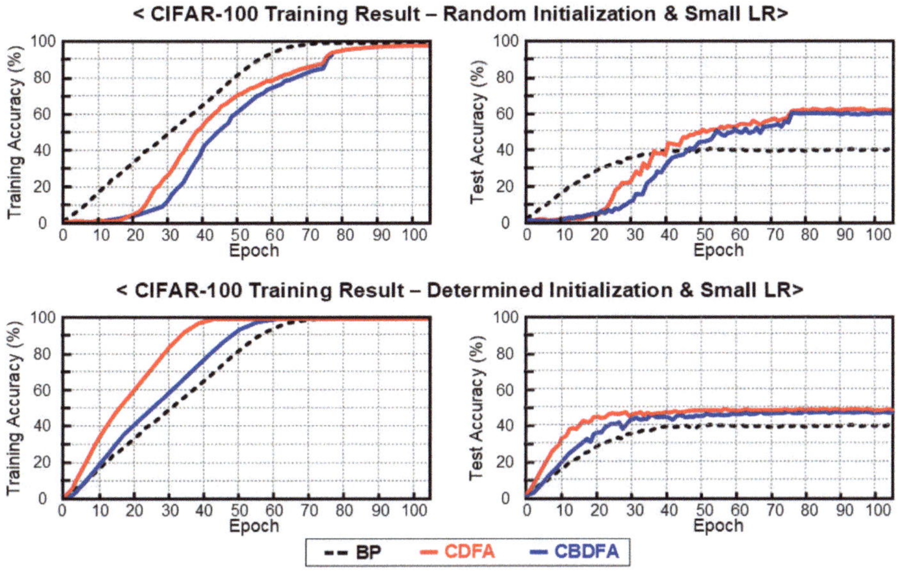

Fig. 3.4 Training and test accuracy with proposed training algorithm

Fig. 3.4. DFA requires time to be adaptive to the randomly initialized feedback weights, so it takes a long latency to be converged. In the BP approach, we generally take the larger learning rate to make the training faster. However, the test accuracy of the DFA and BDFA is swung up and down dramatically when the large learning rate is applied. Moreover, it still spends a long time to converge. In this problem, the initialization with Eqs. 3.7 and 3.9 can be useful to solve the learning speed and stability problem. After the feedback weight is initialized by the proposed equations, it shows faster and more stable convergence characteristic as shown in Fig. 3.5. When the proposed initialization method is combined with the large learning rate, it shows the best training performance compared with the other results.

When the dataset is augmented with the flipping and cropping, the training performance of the BP and CBDFA becomes higher than before. The simulation uses the large learning rate, and CBDFA takes the initialization with Eq. 3.9. In Table 3.2, the performance of the CBDFA shows the highest accuracy compared with not only BP but also the training suggested by Hoffer et al. [102]. It trains only the convolutional layers, and the parameters of the FC layers are fixed. Even considering the data augmentation, CBDFA still shows higher training accuracy compared with the other two methods. As a result, CBDFA seems robust to the size of the dataset. There are some interesting observations in our simulation. First of all, the CBDFA shows negligible accuracy degradation compared with the CDFA based training. Sometimes, the CBDFA has a rather better performance than the CDFA case. Refer to the learning curve described in both Figs. 3.4 and 3.5, the CBDFA's learning speed is slightly degraded, but the final training results are

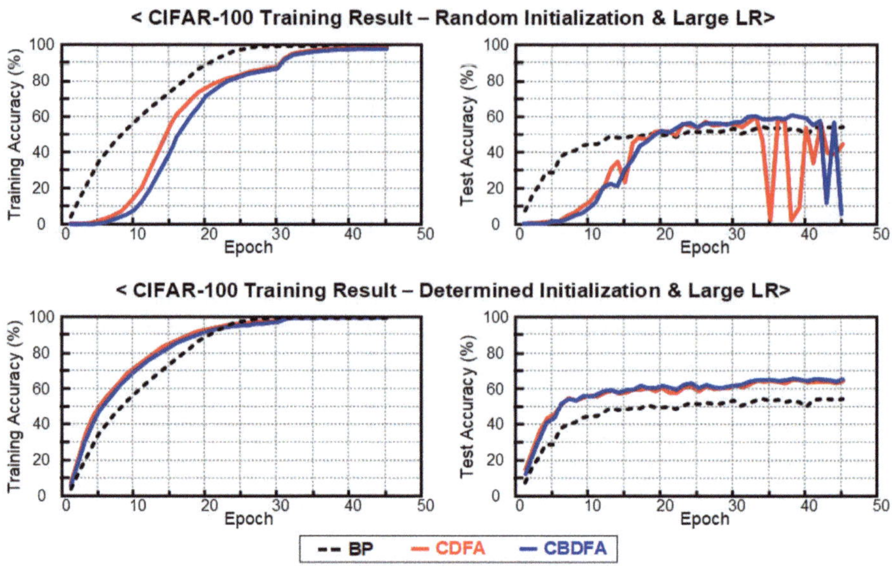

Fig. 3.5 Training and test accuracy with proposed training algorithm

Table 3.2 CNN training result with data augmentation (CIFAR-10)

	W/o data augmentation			W/ data augmentation		
	BP	CBDFA	Conv. only training	BP	CBDFA	Conv. only training
Top5	99.15	99.07	98.84	99.33	99.49	99.46
Top1	82.33	87.35	82.06	87.97	90.13	88.48

approximately the same. Therefore, CBDFA can improve the training performance and take the hardware benefits such as smaller memory bandwidth by adopting binarized feedback weights. The second one is the effect of the BN. Even though the suggested initialization method achieves the fast and stable training, it has a little accuracy degradation compared with the random initialization method. This accuracy degradation can be reduced when the BN layer is added after the FC layer. This result is counter characteristic compared with the BP case because the BP shows the worse training result when the BN layer is followed right after the FC layer. To sum up, Eqs. 3.7 and 3.9 are much more powerful when the BN is followed after the convolutional layer.

3.4.2 Example of Online Learning with Small Dataset

Nam and Han [41] suggested online FC learning based object tracking algorithm, MDNet. The online learning concept in MDNet has derived many other algorithms

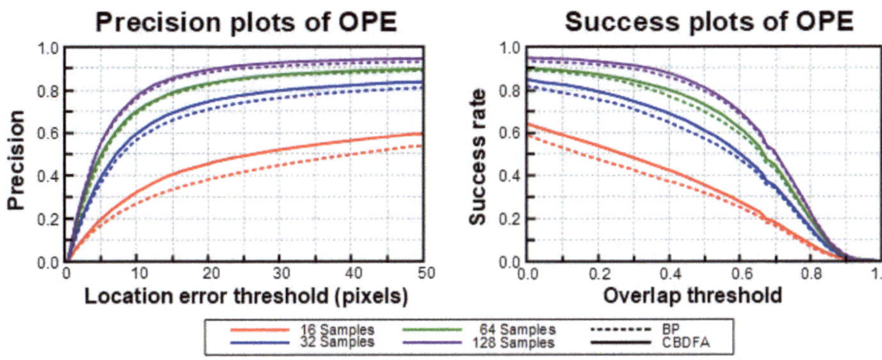

Fig. 3.6 Object tracking result with the proposed training algorithm, CBDFA

such as BranchOut [103] and ADNet [104]. In the MDNet, both the convolutional layers and FC layers are pre-trained with the VOT [105, 106] and OTB [107] object tracking datasets. However, the last layer of the FC layers is randomly initialized for the new tracking task. The convolutional layers do not need to be trained during the tracking, but the FC layers are fine-tuned by using the BP. To apply BDFA to FC online learning with the small dataset, we replace the BP by the BDFA for FC layers in the MDNet.

We compared the object tracking performance by drawing the precision and the success plots of the one-pass evaluation (OPE) in the OTB dataset. As shown in Fig. 3.6, the object tracking with the BDFA based online learning shows similar performance compared with the BP case. However, the BDFA shows better performance than BP when the batch size becomes smaller. Since the object tracking application is very sensitive to the online learning speed and the BDFA has a chance to be computed in parallel, BDFA based online learning becomes much more beneficial than conventional BP. Moreover, the BDFA can dramatically reduce the required data transaction in error propagation because of the fewer neuron interconnections and the binarization. As pointed out in the paper, [43], BP based online learning is inefficient for online learning in the devices that have limited memory bandwidth, computing resources, and small dataset. In this case, the effect of the BDFA can be maximized because of its profits.

3.5 Conclusion

In this chapter, we propose the new DNN learning algorithms to maximize the computation efficiency without accuracy degradation. We adopt one of the training methods, DFA, and combine it with the conventional BP. The combination of the DFA and the BP shows much better test accuracy in CNN training through the simulation in the CIFAFR-10 and CIFAR-100 datasets. BDFA takes one step further, binarizing the feedback weight while maintaining a similar performance compared with the full-precision DFA. The stability problem induced in the DFA and BDFA simulation (Fig. 3.5) can be solved by the new feedback weight initialization method, Eqs. 3.7 and 3.9. The BDFA is also simulated in the object tracking application, and it shows better tracking results compared with the conventional BP based online FC tuning.

In this research, we can see that the DFA can reduce the computational complexity and achieve better training performance than conventional training method. However, the feedback path of the DFA cannot still be applied directly into convolutional layer because of the significant accuracy degradation. To break the limitation of the current research, we will continue the research about expanding the usage of the DFA to not only convolutional layers but also other RNN networks such as LSTM [108] and GRU [109].

Chapter 4
New Algorithm 2: Extension of Direct Feedback Alignment to Convolutional Recurrent Neural Network

Throughout this chapter, we focus on the improvement of the direct feedback alignment (DFA) algorithm and extend the usage of the DFA to convolutional and recurrent neural networks (CNNs and RNNs). Even though the DFA algorithm is biologically plausible and has a potential of high-speed training, it has not been considered as the substitute for back-propagation (BP) due to the low accuracy in the CNN and RNN training. In this chapter, we propose a new DFA algorithm for BP-level accurate CNN and RNN training. First, we divide the network into several modules and apply the DFA algorithm within the module. Second, the DFA with the sparse backward weight is applied. It comes with a form of dilated convolution in the CNN case and in a form of sparse matrix multiplication in the RNN case. Additionally, the error propagation method of CNN becomes simpler through the group convolution. Finally, hybrid DFA increases the accuracy of the CNN and RNN training to the BP-level while taking advantage of the parallelism and hardware efficiency of the DFA algorithm.

This chapter is based on paper "Extension of Direct Feedback Alignment to Convolutional and Recurrent Neural Network for Bio-plausible Deep Learning" (arXiv:2006.12830, [74]).

4.1 Introduction

Recently, deep neural networks (DNNs) have enabled various AI-driven tasks, and the back-propagation (BP) algorithm is recognized as the typical training methodology of the DNNs. However, the BP algorithm is regarded to be biologically implausible for two reasons: (1) back-propagation enforces the backward weight to be always symmetric to the forward weight (weight transport problem, [110]) and (2) the error should be propagated layer-by-layer (backward locking problem, [111]). The weight transport problem induces additional memory transactions

© The Author(s), under exclusive license to Springer Nature Switzerland AG 2023

D. Han, H.-J. Yoo, *On-Chip Training NPU - Algorithm, Architecture and SoC Design*,
https://doi.org/10.1007/978-3-031-34237-0_4

caused by the reading of transposed backward weight at each learning iteration. The backward locking problem makes it hard to parallelize the DNN training since the next feed-forward step cannot proceed until the entire back-propagation step is finished.

Several biologically plausible learning algorithms have been proposed to overcome the limitation of the back-propagation. Target propagation [112] relieved the weight transport problem by using the formulated local loss function instead of the error gradient to update the intermediate layers. Feedback alignment suggested by Lillicrap et al. [44] used the random and fixed backward weight for error propagation instead of transposed forward weight. Nokland [71] solved both the weight transport and backward locking problem by using the direct feedback alignment (DFA). It directly propagates errors from the last output layer to the intermediate layers using the random backward weight. However, those biologically plausible learning algorithms failed to train the convolutional neural networks (CNNs) or recurrent neural networks (RNNs) due to the large accuracy drop.

Han and Yoo [72] showed the CNN can be successfully trained on CIFAR-10 and CIFAR-100 datasets by adopting DFA to the FC layers but back-propagation to the convolutional layers. However, the backward locking problem was partially solved only in the FC layers; thereby, it missed the opportunity of parallelism in the error propagation of convolutional layers.

In this chapter, we suggest new methodologies to improve the accuracy of the DFA algorithm in the CNN and RNN training. The main contributions of our work are summarized as follows:

- First, we divide both the CNN and RNN into several small modules and enable direct error propagation inside the module.
- Second, we modify the DFA algorithm to be applied to the CNN training by using the following features:
  (a) Dilated-convolution-based error propagation is adopted instead of a fully connected manner.
  (b) Error propagation is simplified with the group convolution suggested by Zhang et al. [113].
- Third, we simplify the backward weight used in the RNN training by adopting the upper triangular matrix. Moreover, we combine conventional DFA optimization methods such as binary DFA [72] and sparse DFA [73].
- Fourth, we propose a new training algorithm, hybrid DFA (HDFA), which mixes both BP and DFA based error propagation to achieve BP-level accuracy with the minimum hardware cost.
- Finally, we verified the performance of the proposed method by using the three different network architectures in both CNN and RNN training. Experimental results successfully demonstrate that the proposed algorithm can achieve high training accuracy as same as BP. Moreover, it shows the lowest memory transaction amounts compared with the existing error propagation methodologies.

The remaining part of the chapter is organized as follows. The mathematical notation of the BP and DFA will be explained, and the problem of the BP based algorithm, backward locking, will be introduced in Sect. 4.2. Then, the detailed explanation of the proposed algorithm and the related experiments will be followed in Sects. 4.3 and 4.4. The paper will be concluded in Sect. 4.5, and the detailed results of the experiments will be dealt with in the supplementary materials.

4.2 Background

4.2.1 Back-propagation-based CNN and RNN Training

Back-propagation [40] is a widely used algorithm for the DNN training. It computes gradient of the loss function with respect to the weights of the network by using chain rule. Let us assume that the DNN consists of L layers, and it uses a non-linear function, $f(\cdot)$. The non-linear function should be placed after the multiplication of the input feature map, $\boldsymbol{o}_i$, and corresponding weight, $\boldsymbol{W}_i$. Then, the input of the next layer can be obtained as

$$\boldsymbol{p}_{i+1} = \boldsymbol{W}_i\, \boldsymbol{o}_i, \quad \boldsymbol{o}_{i+1} = f(\boldsymbol{p}_{i+1}), \quad i \in \{0, 1, \ldots, L-1\} \tag{4.1}$$

After loss calculation, the error propagation stage proceeds. In this stage, it propagates the errors from the last layer to the prior layer based on chain rule. The error, $\boldsymbol{e}_{i+1}$, should be generated to make the gradient of the weight, $\boldsymbol{W}_i$. Detail operations of the error propagation and weight gradient stage are described as follows:

$$\boldsymbol{e}_i = (\boldsymbol{W}_i^T\, \boldsymbol{e}_{i+1}) \odot f'(\boldsymbol{p}_{i+1}), \quad i \in \{1, 2, \ldots, L-1\} \tag{4.2}$$

$$\boldsymbol{G}_i = \boldsymbol{e}_{i+1}\, \boldsymbol{o}_i^T, \quad i \in \{0, 1, \ldots, L-1\} \tag{4.3}$$

$$\boldsymbol{W}_i^{new} = \boldsymbol{W}_i - \eta\, \boldsymbol{G}_i \tag{4.4}$$

Equation 4.4 can be varied based on the optimizer such as momentum [100] or adam [114].

In the CNN training, the matrix multiplication operations are substituted with the convolution operations. In addition, 180° flipped kernel is used instead of the transposed weight in Eq. 4.2.

The RNN also adopts BP based training but should consider recurrent computational characteristics. During the RNN training, the RNN is unfolded through time and calculates weight gradients based on BP. Since the unfolded RNN cells have a shared weight, the weight gradient generated in every cell should be integrated to update the corresponding weight. This method is called as "Back-propagation Through Time" (BPTT, [115]).

BP achieved state-of-the-art performance in DNN training, but it currently shows the limitation of parallel computing. As mentioned before, the error propagation should proceed from the last layer to the first layer in order. The next inference can be started after the weight update of the first layer is finished. However, it should wait until the error propagation is finished, and the error reaches the first layer. This problem is generally known as the "backward locking" problem. Nowadays, many researchers tried to multi-GPU for the high-speed DNN training, but the serial computational characteristic appeared in the BP becomes the main obstacle of that purpose.

Specifically, the problem becomes critical in RNN training. Although BPTT shows great training performance compared with other optimization algorithms such as evolution strategy [90], it suffers from a backward locking problem and causes slow training. When the RNN has a large time step, the depth of the unfolded RNN becomes huge. Since the error propagation of the different time steps proceeds with serial manner, the parallel processing is impossible with the BPTT, and the speed of the error propagation can be limited.

4.2.2 Direct Feedback Alignment based CNN and RNN Training

BP based training methodology has two main disadvantages. One of them is the weight transport problem that needs the transpose of the weights during the error propagation. The transpose reading of the weights is not appropriate for biologically plausible training, and it also requires additional memory access cost. Previously, the neuron connections during the error propagation depend on forward neuron connections, but the DFA got rid of this dependency by propagating errors of the last layer directly to the all intermediate layers. The backward weight, $\boldsymbol{D}_i$, consists of random values, but the value should be maintained even the forward weight is newly updated. With the error of the last layer, $\boldsymbol{e}_L$, and related backward weight, $\boldsymbol{D}_i$, the error of the intermediate layer can be calculated as

$$\boldsymbol{e}_i = (\boldsymbol{D}_i\ \boldsymbol{e}_L) \odot f'(\boldsymbol{p}_{i+1}) \tag{4.5}$$

The DFA is one of the biologically plausible training methods, and it has the potential strength to overcome the backward locking that is the main weakness of the BP based training scenario. However, the DFA shows comparable accuracy only in the training of fully connected layers. Han and Yoo [72] tried to enlarge the DFA to CNN, but it shows high accuracy by adopting DFA only to the last few fully connected layers. Since the portion of the fully connected layer becomes smaller and smaller in the current CNN architecture, it is not that useful in the CNNs such as ResNet [116].

In the conventional DFA, the connections of the backward path are considered as the fully connected layer, so the computation methodology of the error propagation is fixed as matrix multiplication. For example, the CNN adopts convolution operations, and RNN adopts recurrent computation flow in the forward path. When we apply the DFA as the error propagation method, it creates new connections between the output neurons and the intermediate neurons. And then, it allocates arbitrary synaptic weights to the newly created backward connections. Therefore, the backward path of CNN and RNN is revised as a fully connected layer, and it performs matrix multiplication instead of convolution or recurrent computation. These different computational characteristics induce low training performance and low hardware efficiency compared with BP. In CNN, the intermediate feature map placed in the front layer generally has a large size. It induces that the DFA has more computation amounts compared with the BP algorithm. When the DFA is applied to the RNN, the problem can be significantly increased due to the time step. The number of outputs can be increased if the size of the time step is large. It induces large memory and computations due to the connections related to the output neurons of every time step.

In summary, conventional DFA was an important step forward to biologically plausible DNN training. However, it cannot be the substitute for using BP because of the low training performance and efficiency.

4.3 Our Approach

In this section, we will introduce the new DFA algorithm that has high scalability to various types of network architecture while maintaining the advantages of the conventional DFA algorithm. At first, we divide the DNN into small modules and apply the DFA inside the module. Second, dilated-convolution- and shufflenet [113]-based channel division will be combined with the DFA algorithm in the CNN training. Third, the upper triangular and highly sparse backward weight is adopted to improve both the accuracy and efficiency of the RNN training. At last, we propose the hybrid DFA (HDFA) that blends BP and DFA to compensate for the accuracy loss caused by DFA.

4.3.1 Network Modularization

As depicted in Fig. 4.1, the DFA based training shows much lower training accuracy compared with the BP algorithm. The main reason for the accuracy drop is a large number of layers. The DFA shows lower accuracy when it is applied to deep CNNs. Han and Yoo [72] show that using DFA only on the last few layers can achieve high

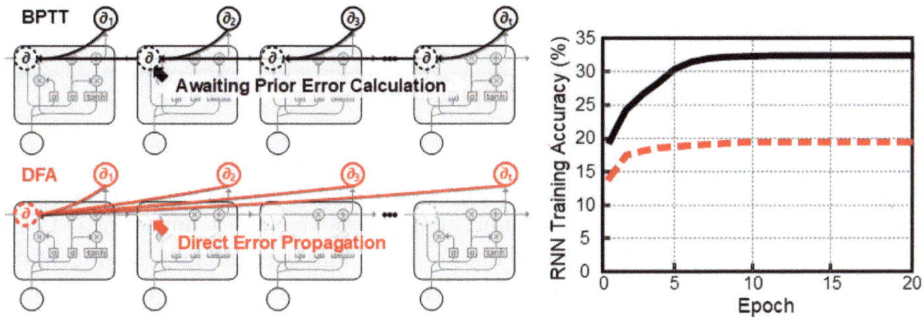

Fig. 4.1 Back-propagation through time vs. DFA based error propagation

training accuracy while maintaining BP based error propagation on the prior layers. Therefore, we tried to revise the original DFA algorithm by reducing the distance of the direct error propagation path.

Our proposed DFA adopts modularization that divides the network into several modules. The distance of direct connection becomes shorter than before because the direct error propagation is done only inside the module. The number of modules is fixed in RNN training, but it is a design choice in the CNN training.

In the CNN training, the number of modules has a trade-off between training accuracy and a chance of parallel processing. When the number of modules is increased, the training accuracy of the DFA can be increased because the distance of the error propagation becomes shorter and the computation characteristic becomes similar to BP. In contrast, it can disturb the multi-GPU acceleration because the number of direct paths is decreased due to modularization. Let M be the total number of error propagation module; then the computation of DFA is revised as follows:

$$\begin{cases} \boldsymbol{e}_{i,j} &= (\boldsymbol{D}_{i,j}\ \boldsymbol{e}_{i,L_i}) \odot f'(\boldsymbol{p}_{i,j+1}) \quad j \neq -1 \\ \boldsymbol{e}_{i-1,L_{i-1}} &= (\boldsymbol{D}_{i,j}\ \boldsymbol{e}_{i,L_i}) \odot f'(\boldsymbol{p}_{i,0}) \quad\ \ j = -1 \end{cases} \quad i \in \{1, 2, \dots, M\}, \quad j \in \{-1, 0, 1, \dots, L_i - 1\} \tag{4.6}$$

The number of layers is small in the RNN cells so we divide the RNN cell only into two modules. The first module includes both the embedding layer and the basic RNN cell. The other module contains the last fully connected layer that decides the final output based on the result of RNN cells. The original DFA based RNN training suffers from a large computation amount because there are lots of output neurons in the last layer. Thanks to modularization, direct propagation is only performed from the RNN cell that has a much smaller number of output neurons. Moreover, this division induces accuracy improvement due to the short error propagation path.

4.3.2 Computational Optimization of CNN

Dilated-Convolution-Based Error Propagation

Training CNN with the conventional DFA requires error propagation in a fully connected fashion. However, as seen in Fig. 4.2a, a single output element is affected only by the elements inside the specific area, receptive field. Conventional DFA does not consider the receptive field but connects massive unrelated connections during the error propagation. As a result, the complexity of error propagation grows exponentially and makes it hard to train the network with the conventional DFA. In our proposed DFA, these unnecessary connections are removed by adopting the convolutional direct error propagation. As shown in Fig. 4.2b, error in the convolution layer is propagated by the series of convolutions during the back-propagation. In consequence, the error of the $L + 2$th layer requires a widened receptive field to generate a single error element at layer L. In order to directly propagate error from layer $L+2$ to L, the same receptive field should be maintained.

Yu and Koltun [117] adopted dilated-convolution-based network architecture to exploit a large receptive field without increasing the size of the forward weight. The additional computation required by using an expanded receptive field can be resolved by utilizing the zero-skipping DNN hardware [118]. Therefore, we adopted the dilated convolution for direct error propagation that occurred within the module. As shown in Fig. 4.2b, the convolutional DFA enlarges the receptive field of the backward weight to the required size by inserting zeros inside this backward weight. In addition to this, binary backward weight is utilized to minimize the size of the required memory to store the backward weight. In this way, the error is propagated efficiently in terms of both memory and computation while maintaining its alignment with the forward propagation.

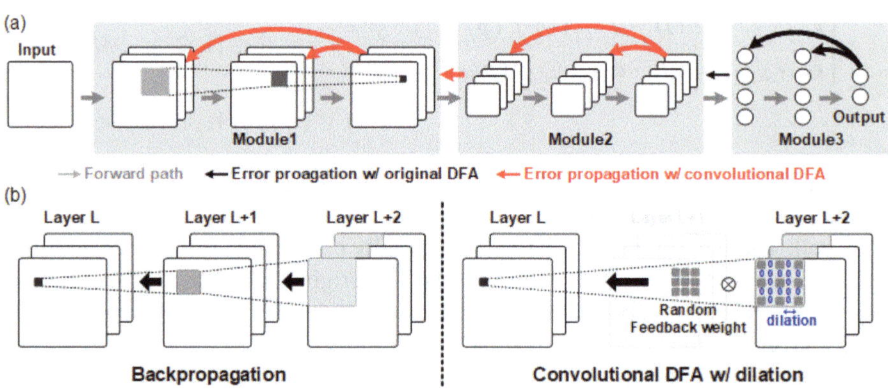

Fig. 4.2 (**a**) Error propagation path by using DFA with network modularization. (**b**) Training the convolutional layer by using back-propagation (left) and convolutional DFA (right)

Division of Channels During Error Propagation

Error propagation with dilated convolution makes the computation and memory requirement comparable to that of the back-propagation. In this chapter, we further optimized the error propagation computation by adopting the group convolution and channel shuffling concept used in the previous mobile-optimized network, ShuffleNet [113].

During the group-convolution-based error propagation, the errors are divided into several groups of channels, and each group is computed independently. Denote the error as $\boldsymbol{e}_i = \left[\boldsymbol{e}_i^1, \boldsymbol{e}_i^2, \cdots, \boldsymbol{e}_i^G\right]$, where $i \in \{1, 2, \cdots, L-1\}$ represents the layer number. $\boldsymbol{e}_i^j$ denotes error at jth channel group, where $j \in \{1, 2, \cdots, G\}$ represents the channel group number. With a $k \times k$ backward weight $\boldsymbol{D_1}$ and 1×1 backward weight $\boldsymbol{D_2}$, calculation of the error $\boldsymbol{e}_i$ is formulated as below.

$$\boldsymbol{e}_i = (\boldsymbol{D_1} \otimes^g shuffle(\boldsymbol{D_2} \otimes^g \boldsymbol{e}_L)) \odot f'(\boldsymbol{p}_{i+1}) \tag{4.7}$$

Group convolution between weight, $\boldsymbol{D}$, and error, $\boldsymbol{e}$, can be expressed as follows:

$$\boldsymbol{D} \otimes^g \boldsymbol{e} = \left[\boldsymbol{D}^1 \otimes \boldsymbol{e}^1, \boldsymbol{D}^2 \otimes \boldsymbol{e}^2, \ldots, \boldsymbol{D}^G \otimes \boldsymbol{e}^G\right] \tag{4.8}$$

Compared with the standard convolution $\boldsymbol{D} \otimes \boldsymbol{e}$, $\boldsymbol{D}^i \otimes \boldsymbol{e}^i$ requires $1/G^2$ times less computation by removing channel connections among groups. Therefore, group convolution involves $1/G$ less number of computation. Likewise, the size of the backward weight, $\boldsymbol{D}$, is $1/G$ times smaller thanks to the channel division.

As a result of modified error propagation methodology, both memory transaction and the number of computation are reduced as follows. Denote {width, height, and number of channels} for $\boldsymbol{e}_L$ and $\boldsymbol{e}_i$ as $\{w_L, h_L, ch_L\}$ and $\{w_i, h_i, ch_+i\}$, respectively. Convolutional DFA used in Sect. 4.3.2 requires $k^2 \times ch_L \times w_i \times h_i \times ch_i$ multiplications. On the other hand, proposed group-convolution-based error propagation requires $k^2 \times ch_L \times w_i \times h_i \times ch_i \times (1/k^2 + ch_L/(ch_i \times G))$ multiplications. For instance, assume $1/k^2 \ll 1/G$ and $ch_i = ch_L$. Then the low complexity error propagation method approximately reduces the memory and computation requirement by G times (Fig. 4.3).

4.3.3 Computational Optimization of RNN

Revision of Backward Weight into Upper Triangular Matrix

Refer to [44], the angle between backward weight, $\boldsymbol{D}$, and the transpose of forward weight, $\boldsymbol{W}^T$, becomes similar when the training proceeds. In other words, the network automatically adapts its random backward weights by modifying the corresponding forward weights. Therefore, the training accuracy can be dropped

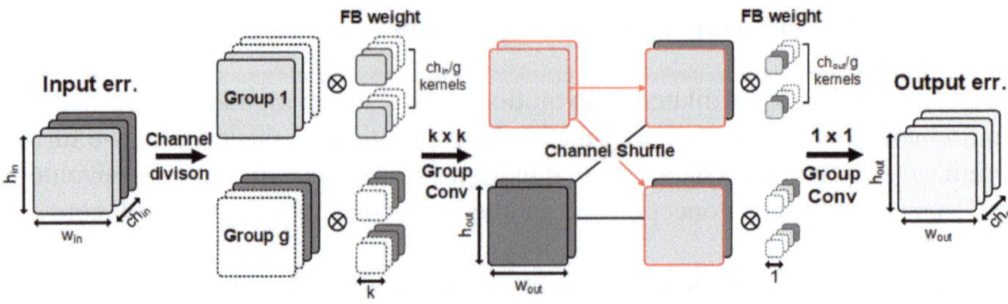

Fig. 4.3 Group-convolution-based error propagation method for CNN

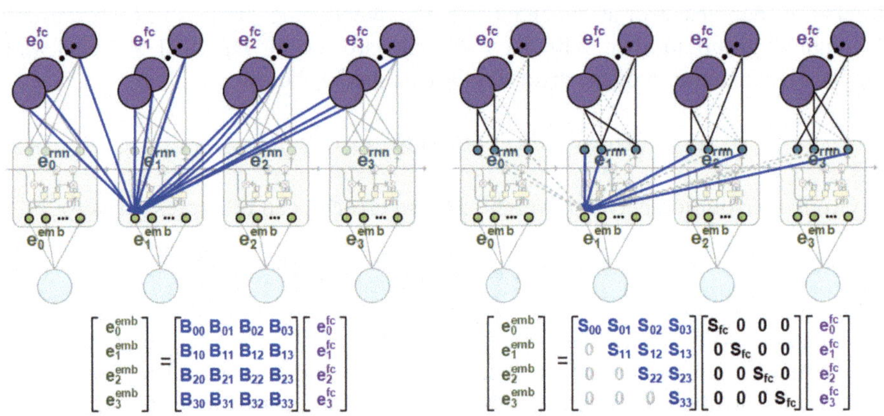

Fig. 4.4 Proposed RNN training method by using DFA (left: original DFA, right: modularization + upper triangular backward weight + sparse backward weight)

when the forward weight cannot easily predict the backward weight during the training process.

In the conventional DFA based RNN training, all intermediate neurons are connected with the output neurons of the last layer during the error propagation. However, this learning process makes the training process more difficult because of the unnecessary backward connections. Let the $\boldsymbol{X}_i$ be the input value of the RNN cell at the time step i. Figure 4.4 shows the unfolded form of the RNN computation. Final output at the time step $\boldsymbol{i}$ is affected by the inputs, $\boldsymbol{X}_0$ to $\boldsymbol{X}_{i-1}$, which are the values of the previous and current inputs. Since the backward connection is constructed even with the future inputs ($\boldsymbol{X}_{i+1}, \boldsymbol{X}_{i+2}, \ldots$) in the conventional DFA, it finally becomes the main reason for the low accuracy.

To improve the performance of the DFA, we disconnect the backward connections that have no impact on the final output at the forward computation. The backward weight used in the DFA is modified to reflect the data dependency, and the backward weight shows the shape of the upper triangular matrix as depicted in Fig. 4.4. Another observation is that the input values of RNN cells experience the same operations even in the next time step. The backward weight is modified to reflect the repeated operation by using the power of partial backward weight. In summary, the final backward weight is determined as follows when the number of output neurons of the RNN cell is H and the total time step is T.

$$\boldsymbol{B}^{rnn}_{HT\times HT} = \begin{cases} \boldsymbol{B}^{rnn}_{i,j} = power(\boldsymbol{D}^{rnn}_{i,j}, T - j) & i \le j \\ \boldsymbol{B}^{rnn}_{i,j} = \boldsymbol{O}_{H\times H} & i > j \end{cases} \quad i, j \in \{0, 1, \ldots, T-1\} \tag{4.9}$$

Highly Sparse and Binary Backward Weight

The backward weight used in the RNN training becomes huge because the backward connections are constructed for every output neuron of unfolded RNN cells. These connections induce the large computation amount and storage to store the backward weight. One of the possible solutions is the binarization introduced by [72]. It modifies the backward weight that has only the sign values, +1/−1. This modification shows no accuracy degradation but reduces the total memory storage by 96.9%. It successfully reduces the memory size but does not affect the total number of computations.

The other solution is using the sparse backward weight suggested by Crafton et al. [73]. Crafton et al. [73] verified that the performance of the DFA is not that degraded with the backward weight including lots of zeros. The sparse weight can reduce the overall necessary computations during the error propagation. Zero-skipping DNN accelerator architecture such as EIE [45] can be efficiently used to improve the processing speed of the error propagation stage. We finally combine both the binarized and highly sparse backward weight to optimize both memory storage and computations during the RNN training. The effect of the binarization and sparse backward weight will be discussed in Sect. 4.4. When the sparse bitmap mask is denoted as R, the final backward weight is calculated as follows:

$$\boldsymbol{S}^{fc} = sign(\boldsymbol{D}^{fc} \odot \boldsymbol{R}^{fc}), \quad \boldsymbol{e}^{rnn}_{i} = (\boldsymbol{S}^{fc}\, \boldsymbol{e}^{fc}_{i}) \odot f'(\boldsymbol{p}^{fc}) \tag{4.10}$$

$$\boldsymbol{S}^{rnn}_{HT\times HT} = \begin{cases} \boldsymbol{S}^{rnn}_{i,j} = sign(power(\boldsymbol{D}^{rnn}_{i,j}, T - j) \odot \boldsymbol{R}^{rnn}_{j}) & i \le j \\ \boldsymbol{S}^{rnn}_{i,j} = \boldsymbol{O}_{H\times H} & i > j \end{cases}, i, j \in \{0, 1, \ldots, T-1\} \tag{4.11}$$

$$\boldsymbol{e}^{emb} = \begin{bmatrix} \boldsymbol{e}_0^{emb} \\ \boldsymbol{e}_1^{emb} \\ \vdots \\ \boldsymbol{e}_{T-1}^{emb} \end{bmatrix}, \quad \boldsymbol{e}^{rnn} = \begin{bmatrix} \boldsymbol{e}_0^{rnn} \\ \boldsymbol{e}_1^{rnn} \\ \vdots \\ \boldsymbol{e}_{T-1}^{rnn} \end{bmatrix}, \quad \boldsymbol{e}^{emb} = (\boldsymbol{S}^{rnn}\ \boldsymbol{e}^{rnn}) \odot f'(\boldsymbol{p}^{rnn}) \tag{4.12}$$

4.3.4 Hybrid Direct Feedback Alignment

Algorithm 1 Custom optimizer of the hybrid direct feedback alignment

Input: Learning rate η, momentum parameter α, mix parameter γ, BP ratio p
Input: Initial parameter $\boldsymbol{\theta}$, Initial BP velocity $\boldsymbol{v}^{BP}$, Initial DFA velocity $\boldsymbol{v}^{DFA}$
Input: Objective function $\boldsymbol{f}(\boldsymbol{x};\boldsymbol{\theta})$ given input $\boldsymbol{x}$ and parameters $\boldsymbol{\theta}$
  Initialize network parameters $\boldsymbol{\theta}$
  Initialize momentum for BP $\boldsymbol{v}^{BP} \leftarrow 0$ and momentum for DFA $\boldsymbol{v}^{DFA} \leftarrow 0$
  while $\boldsymbol{\theta}$ not converged **do**
    Sample a minibatch of m examples from the training set $\{x^{(1)}, \cdots, x^{(m)}\}$ with corresponding targets $y^{(i)}$
    Set $rand = random(0,\ 1)$
    if $rand \le p$ **then**
      $\boldsymbol{g} \leftarrow \frac{1}{m}\nabla_{\boldsymbol{\theta}}^{BP} \sum_i L(f(\boldsymbol{x}^{(i)}, \boldsymbol{\theta}), \boldsymbol{y}^{(i)})$ ▷ Compute gradient through BP
      $\boldsymbol{v}^{BP} \leftarrow \boldsymbol{v}^{BP} + \alpha \boldsymbol{g}$ ▷ Compute momentum update
      $\boldsymbol{\theta} \leftarrow \boldsymbol{\theta} + \eta \boldsymbol{v}^{BP}$ ▷ Update parameter
    else
      $\boldsymbol{g} \leftarrow \frac{1}{m}\nabla_{\boldsymbol{\theta}}^{DFA} \sum_i L(f(\boldsymbol{x}^{(i)}, \boldsymbol{\theta}), \boldsymbol{y}^{(i)})$ ▷ Compute gradient through DFA in Sect. 4.3.2
      $\boldsymbol{v}^{DFA} \leftarrow \boldsymbol{v}^{DFA} + \alpha \boldsymbol{g}$ ▷ Compute momentum update
      $\boldsymbol{u} \leftarrow \gamma \boldsymbol{v}^{DFA} + (1-\gamma)\boldsymbol{v}^{BP}$ ▷ Mix DFA momentum with BP momentum
      $\boldsymbol{\theta} \leftarrow \boldsymbol{\theta} + \eta \boldsymbol{u}$ ▷ Update parameter
    end if
  end while
  return $\boldsymbol{\theta}$

We can get higher accuracy by modifying the error propagation methods as explained in Sects. 4.3.1–4.3.3. However, the accuracy of the DFA in both CNN and RNN is still below the one trained with the BP algorithm. Hybrid DFA (HDFA) is newly proposed to be comparable with the BP algorithm in terms of the training accuracy.

In the case of RNN training, the HDFA works as follows. The error propagation method is randomly determined to be BP or DFA at each iteration. The weight gradients generated by two different error propagation methods are integrated into the momentum term of Adam optimizer [114]. We found that the gradient occasionally generated by BP accelerates the convergence of the DFA algorithm, and it finally helps the DFA algorithm to have higher accuracy.

The optimizer used in CNN training is different from conventional optimizers such as momentum or Adam. The HDFA based CNN training with the conventional optimizers fails to accomplish BP-level accuracy. To train the CNN with the proposed HDFA, original momentum optimizer [100] should be modified as described in Algorithm 1. Unlike the conventional optimizers, our proposed optimizer distinguishes the momentum terms of the DFA from the BP. We denote the momentum of the BP as $\boldsymbol{v}^{BP}$ and the DFA as $\boldsymbol{v}^{DFA}$. The error generated by each algorithm affects only its own momentum value. The frequency of the BP algorithm is determined by the hyperparameter, BP ratio (p). When the error is propagated through the DFA, both $\boldsymbol{v}^{DFA}$ and $\boldsymbol{v}^{BP}$ are considered to make the final weight gradient. Thanks to the revision of optimizer, the HDFA can achieve high training accuracy even in the CNN training scenario.

4.4 Experiments

In this section, we compared the training accuracy of the conventional BP, DFA, and the proposed new DFA algorithm, HDFA. In the CNN training scenario, we used the CIFAR-10 dataset [99] with the random cropping and random horizontal flip data augmentation method. We verified the performance in the various CNNs such as VGG16 [94] and ResNet [116]. The simulation was based on mini-batch gradient descent with the batch size, 128. The gradient optimizer is based on momentum [100], but it is revised based on Algorithm 1.

In the RNN training, the WikiText-2 dataset [119] is used to predict the English word sequence. We tested three different types of RNN cells: (1) first of all, vanilla RNN that only contains the tanh activation function, (2) long short-term memory (LSTM, [108]), and (3) gated recurrent unit (GRU, [109]). All RNN cells receive 200 embedding values and extract 200 values of hidden state, while it adopts a 35-time step and 20 batch size. The optimizer of the RNN is decided as Adam [114] because it shows a higher accuracy compared with the momentum.

4.4.1 CNN Training Result

The training performance of the conventional and proposed algorithm is summarized in Table 4.1. In conventional DFA based training, it shows poor accuracy when the number of layers is increased. The network modularization helps the DFA to have higher accuracy since it reduces the distance of the direct error propagation path. The modification of the error propagation methodology affects training performance in different ways. In the VGG-16, the training accuracy is getting lower when the error propagation method becomes simpler due to the dilated and group convolutions. On the other hand, the modifications make the ResNets have higher accuracy. In this experiment, we verified that the residual connection of

Table 4.1 CNN training and test accuracy [%]

	Channel groups	Network					
		VGG16		R-18		R-50	
		Train	Test	Train	Test	Train	Test
BP	1	99.91	92.68	99.94	92.74	99.97	93.07
Original DFA	–	86.66	77.78	66.53	66.19	68.95	71.92
Convolutional DFA	1	84.84	78.18	81.79	76.55	–	–
+Modularization	1	84.96	78.56	88.99	79.60	82.81	77.99
+Dilation	1	81.22	77.41	88.05	78.09	82.44	77.27
+Group convolution & channel shuffling	4	82.60	76.85	88.99	79.09	86.57	80.4
	8	80.75	77.5	87.94	79.03	87.17	81.36
+BP/BDFA = 0.1/0.9	4	93.39	87.92	94.05	88.35	94.37	88.55
	8	93.64	87.98	93.39	88.29	93.71	87.93
+BP/BDFA = 0.3/0.7	4	97.45	90.15	97.66	91.10	97.41	90.13
	8	97.45	90.28	97.60	90.83	98.05	91.51
+BP/BDFA = 0.5/0.5	4	98.90	91.10	99.18	91.78	98.87	91.42
	8	99.00	91.46	99.27	92.18	99.57	92.21

the ResNet helps the robustness to the DFA based training and shows the best result when it is combined with the simpler error calculations. Although VGG-16 shows lower accuracy compared with the original DFA, the accuracy degradation can be compensated with the hybrid training algorithms. We examined three different BP/DFA ratios by adopting HDFA introduced in Sect. 4.3.4. The HDFA finally raises accuracy as similar to the final result of BP.

Thanks to the modification of error calculation and HDFA, we can get the BP-level high accuracy with less hardware cost. Table 4.2 shows the size of the memory transactions and the number of operations required during the error propagation stage. Since the number of the class is small in the CIFAR-10 dataset, both conventional and proposed DFA algorithms show a smaller number of operations during the error propagation stage. (The detailed relationship between class number and operation will be discussed in Appendix A.1.) Instead, conventional DFA requires large backward weight memory transactions. That is because BP shares weights for multiple channels, but DFA allocates different weights for all connections. Dilated- and group-convolution-based error propagation solve this problem, and finally, the proposed DFA algorithm shows 99.0% lower memory transaction of backward weight data and 60.1% smaller number of operations compared with the BP.

4.4.2 RNN Training Result

Table 4.3 summarizes the overall training performance of the conventional and proposed algorithms. Conventional DFA shows the worst accuracy, but the per-

Table 4.2 Comparison of hardware efficiency in the CNN training

(a) Memory transaction during EP [MB]

	Channel groups	Network		
		VGG16	R-18	R-50
BP	1	56.13	42.64	53.5
DFA	–	6.97	25.39	149.83
Revised DFA	4	0.54	0.43	1.06
	8	0.27	0.23	0.54
+BP/BDFA = 0.1/0.9	4	6.10	4.65	6.30
	8	5.87	4.47	5.84
+BP/BDFA = 0.3/0.7	4	17.22	13.09	16.79
	8	17.03	12.95	16.43

(b) The number of operations during EP [GOP]

	Channel groups	Network		
		VGG16	R-18	R-50
BP	1	0.62	1.48	1.53
DFA	–	0.003	0.01	0.08
Revised DFA	4	0.26	0.56	1.17
	8	0.13	0.31	0.61
+BP/BDFA = 0.1/0.9	4	0.3	0.65	1.2
	8	0.18	0.42	0.7
+BP/BDFA = 0.3/0.7	4	0.37	0.84	1.28
	8	0.28	0.66	0.88

formance of the DFA is improved with the upper triangular backward weight and backward weight initialization suggested by Han and Yoo [72]. The revision of the DFA algorithm increases the training and test accuracy by a maximum of 5.2%p and 3.2%p. The performance of the DFA is further improved when the DFA is blended with BP. As increasing the ratio of BP, we can compensate for the accuracy degradation and achieve similar test accuracy compared with BP.

One of the key observations is that test accuracy appeared in the HDFA is similar to the BP even though the training accuracy still appears to be lower. This observation is the same phenomenon shown in the previous DFA training paper, [72] and it shows that the DFA has the potential of overfitting prevention. When the sparsity of the backward weight is increased, the training accuracy is decreased, while it maintains the test accuracy.

Table 4.4 shows the hardware efficiency of the DFA algorithm during the RNN training. We measure the predicted memory transaction amount and the number of operations appeared during the error propagation stage. Since the amount of memory transaction can be varied depending on its encoding and decoding methods, we calculate the memory transaction by counting the number of nonzero values. We assume that the binarization does not affect the total number of operations and count only the nonzero values as the required computation amount. Upper triangular

Table 4.3 RNN training and test accuracy [%]

	BW	Network					
		RNN		LSTM		GRU	
	Sparsity	Train	Test	Train	Test	Train	Test
BP	–	29.55	23.00	32.24	24.48	32.53	24.52
Original DFA	–	20.69	18.35	20.84	19.19	19.62	18.15
+ Triangular BW	–	22.93	20.33	23.30	20.70	23.04	20.65
+ BW Init	–	24.30	21.00	24.54	21.19	24.77	21.35
+BP/DFA*	0%	27.69	22.48	29.50	23.62	30.54	23.70
=0.1/0.9	80%	27.08	22.66	29.28	24.03	30.04	24.16
	95%	26.85	22.47	28.77	23.99	29.84	24.18
	98%	26.16	22.41	28.61	24.04	29.88	24.25
+BP/DFA*	0%	28.30	22.77	30.65	24.11	31.28	24.18
=0.2/0.8	80%	28.02	22.79	30.57	24.33	30.95	24.39
	95%	27.27	22.73	30.23	24.36	31.10	24.41
	98%	26.50	22.71	30.42	24.32	31.02	24.46
+BP/BDFA*	0%	27.75	22.55	29.70	23.70	30.02	23.80
=0.1/0.9	80%	27.02	22.59	29.50	23.98	30.60	24.18
	95%	26.84	22.48	29.30	24.11	30.28	24.23
	98%	26.12	22.42	28.10	23.97	30.19	24.23
+BP/BDFA*	0%	28.08	22.83	30.77	24.18	31.10	24.22
=0.2/0.8	80%	27.34	22.79	30.07	24.25	30.36	24.42
	95%	27.66	22.76	30.26	24.27	30.97	24.46
	98%	26.84	22.63	29.89	24.37	31.14	24.41

backward weight increases the portion of zero values by disconnecting connections between outputs and unrelated input neurons. It reduces the overall memory transaction and the number of computations by 44.6% compared to the conventional DFA. Thanks to the backward weight binarization, we can compress the weight significantly while maintaining its training performance. After considering the sparse backward weight, both memory transactions and the nonzero operations become smaller compared with the BP approach. When the 98% sparsity is used for backward weight, it shows 87.9% lower memory transaction and the number of operations compared with the BP.

4.4.3 Summary of the HDFA based DNN Training

According to the CNN and RNN training experiments, we verified that the proposed HDFA training algorithm can achieve BP-level training performance but shows the lowest hardware cost to accelerate the error propagation stage. Figure 4.2 shows a summary of the training performance by considering both accuracy and the hardware cost. As shown in Fig. 4.2, the conventional DFA algorithm shows low

Table 4.4 Comparison of hardware efficiency in the RNN training

(a) Memory transaction during EP [MB]

	BW	Network		
	Sparsity	RNN	LSTM	GRU
BP	–	26.94	27.90	27.58
DFA	–	1863.6	1863.6	1863.6
Revised DFA	–	1032.6	1032.6	1032.6
+BP/BDFA	0%	31.72	31.82	31.79
=0.1/0.9	80%	8.49	8.59	8.56
	95%	4.14	4.23	4.20
	98%	3.27	3.36	3.33
+BP/BDFA	0%	31.19	31.38	31.32
=0.2/0.8	80%	10.54	10.74	10.67
	95%	6.67	6.86	6.80
	98%	5.90	6.09	6.02

(b) The number of operations during EP [GOP]

	BW	Network		
	Sparsity	RNN	LSTM	GRU
BP	–	9.4	9.7	9.6
DFA	–	18.6	18.6	18.6
Revised DFA	–	10.3	10.3	10.3
+BP/BDFA	0%	10.2	10.2	10.2
=0.1/0.9	80%	2.7	2.8	2.8
	95%	1.4	1.4	1.4
	98%	1.1	1.1	1.1
+BP/BDFA	0%	10.1	10.1	10.1
=0.2/0.8	80%	3.5	3.5	3.5
	95%	2.2	2.3	2.3
	98%	2.0	2.1	2.0

accuracy even with the high hardware cost, so it is not considered as the alternative to BP. With the revision of the DFA algorithm, we can remove the unnecessary connections and improve the accuracy compared with the conventional algorithms, but it still has the low test accuracy. HDFA with binarized and sparse backward weight finally achieves BP-level test accuracy with the lowest memory transaction and computations.

Proposed DFA is the trial of biologically plausible DNN training and achieves high accuracy as same as the BP algorithm. Moreover, it shows fewer computation amounts compared with the conventional error propagation methodology. Since the DFA is free from the backward locking problem, it also has an advantage of parallel processing. In the CNN training, we can compute errors of multiple layers in parallel. In a single layer, the error propagation can be computed in parallel because the group convolution breaks the data dependency that occurred by the channel. In summary, the proposed DFA enables both layer-level and channel-level parallel processing by using multi-GPU.

In the RNN training, the advantage of the DFA can be magnified. In the BPTT-based training flow, the error propagation suffers from a backward locking problem because the errors of the prior time step can be calculated only after the posterior error calculation is completed. The DFA directly calculates all time-step errors at once, so the error propagation can be accelerated with the multi-GPU processing. In our experiment setup, error propagation speed can be improved 35 × faster than BP even without considering zeros that appeared in the backward weights.

4.5 Conclusion

In this chapter, we develop the new DFA algorithm to be applied to the two general DNNs: CNN and RNN. At first, we divide the network into small fractional modules and connect the direct error propagation path only inside the module. We apply the dilated convolution and channel division to optimize the error propagation of the CNN. In the RNN training scenario, we disconnect unnecessary connections that disturb the accurate training and minimize the backward weight complexity for the low hardware cost. Finally, we suggest HDFA, which mixes both BP and DFA based training method in order to achieve high accuracy as same as the conventional BP. Thanks to the synergy of the hybrid training method, our proposed training method is biologically plausible and also achieves BP-level training performance with the lower hardware cost. The proposed algorithm has the potential of energy-efficient DNN learning acceleration with the application-specific integrated circuits such as [120]. In addition, our proposed method solved the backward locking problem that appeared during the error propagation stage, so the acceleration of error propagation can be possible with the multi-GPU processing.

In this research, we focused on the training of CNN and RNN from scratch. We will continue the research about the potential of the DFA algorithm in the distributed learning scenarios. For example, fine-tuning or federated learning [121] can be examples of our application. Since we verified that the DFA algorithm shows lower hardware cost compared with the BP algorithm, we will develop the DFA algorithm for the light fine-tuning or federated learning methodology.

4.6 Appendices

4.6.1 Direct Feedback Alignment Based Convolutional Neural Network Training

Synergy of the DFA with BP

Table 4.3 shows the learning curve to demonstrate the effectiveness of HDFA. It is clear that using only DFA to train the network degrades the network performance critically. However, by combining the DFA with BP, this accuracy degradation can be compensated. To further verify that using BP and DFA together does not interrupt each other, we conducted an additional experiment. We design the BP only training scenario that assumes the weight gradient of the DFA is ignored to update the new weight. In this scenario, BP is occasionally performed to update the network, and weight is not updated when DFA is applied. As shown in Table 4.3, the HDFA shows a faster and higher learning curve compared with the BP only cases. It concludes that the HDFA can achieve accuracy comparable with BP thanks to the synergy of the two different learning algorithms.

Scalability Issue of the DFA

DFA proposed by [71] suffers from the feedback weight scalability issue on the CNN network. From Fig. 4.4, which compares the feedback weight ratio between the original DFA and our proposed DFA, two observations can be found. First, the proposed DFA is more memory efficient in convolution layers at the front. Second, proposed DFA becomes more beneficial as the CNN network becomes deeper. Motivated by this finding, a detailed analysis of feedback weight scalability is studied as follows.

Denote {width, height, channel} of output error $\boldsymbol{e}_i$ as $\{w, h, ich\}$, kernel size as k, the number of channels at the input error and the number of output classes as och, $class$, respectively. The number of feedback weights for the original DFA is expressed as $class \times w \times h \times ich$. The number of feedback weights for DFA introduced in Sect. 4.3.2 is $k \times k \times och \times ich \times 1/G$, where G is the number of groups used in channel division. As a result, the ratio of feedback weight $\boldsymbol{D}$ used in two different DFA methods is determined by the below equation.

$$ratio(\boldsymbol{D}_{ori}^{DFA} / \boldsymbol{D}_{proposed}^{DFA}) = (w \times h)/(k \times k) \times (class/och) \tag{4.13}$$

Figure 4.5 implies that the original DFA becomes inefficient as the input feature size and the number of the output class scales due to Eq. 4.13.

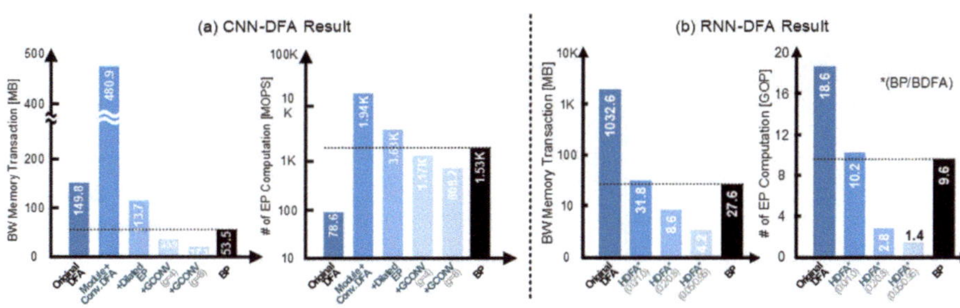

Fig. 4.5 Comparison of hardware efficiency (**a**) CNN-DFA: ResNet50, (**b**) RNN-DFA: GRU

4.6.2 *Direct Feedback Alignment Based Recurrent Neural Network Training*

Synergy of the DFA with BP

Since the DFA has a different backward weight compared with the BP, it can be the obstacle of the training. We did the experiment to prove that the DFA and BP create a synergy effect. Figure 4.6 shows the results of the experiment. In this experiment, we compare the training performance of the HDFA and the BP only training scenarios. In the HDFA scenario, only a certain portion of iterations are trained with the BP algorithm, but the other portion is done by DFA. In contrast, the BP only training scenario performs only the BP based weight update. As shown in Fig. 4.6, HDFA shows faster and higher training curve compared with the BP only performed. Furthermore, HDFA shows even higher performance compared with the DFA only training scenario. This curve can prove that the combination of the DFA and BP algorithms can achieve BP-level training accuracy with the low hardware cost.

Backward Weight Initialization Method

[72] proposed a backward weight initialization method that is only applicable for fully connected layers. It determines backward weight by assuming that all non-linear functions are removed. We also adopted the backward weight initialization concept for RNN training. Since we divide the last decoding fully connected layers

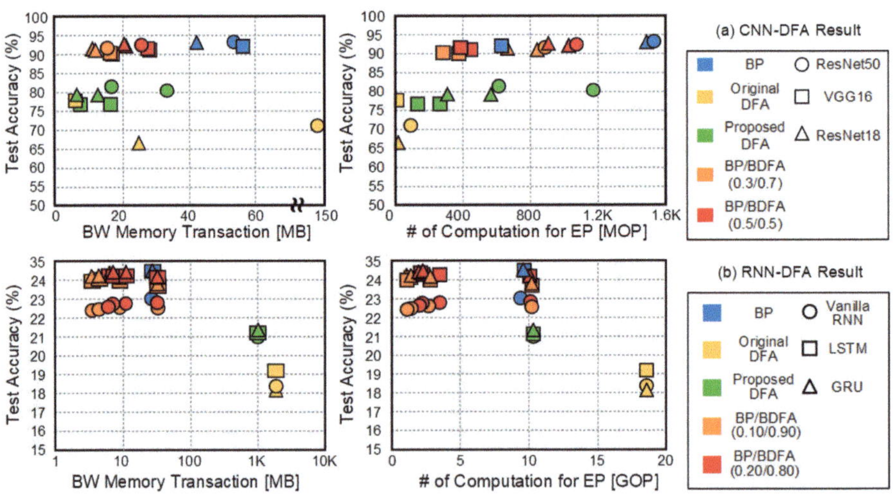

Fig. 4.6 Performance summary of the proposed algorithm: (**a**) CNN-DFA, (**b**) RNN-DFA

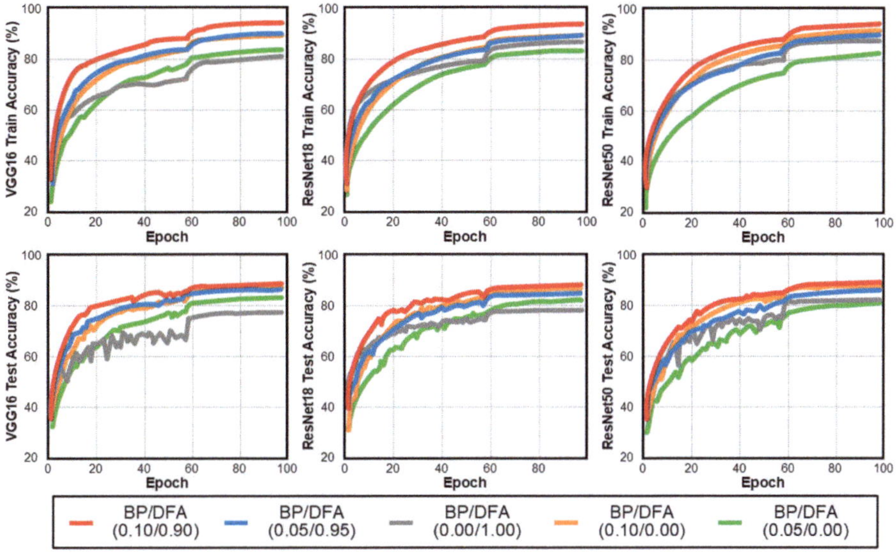

Fig. 4.7 CNN training and test accuracy curve with the proposed HDFA

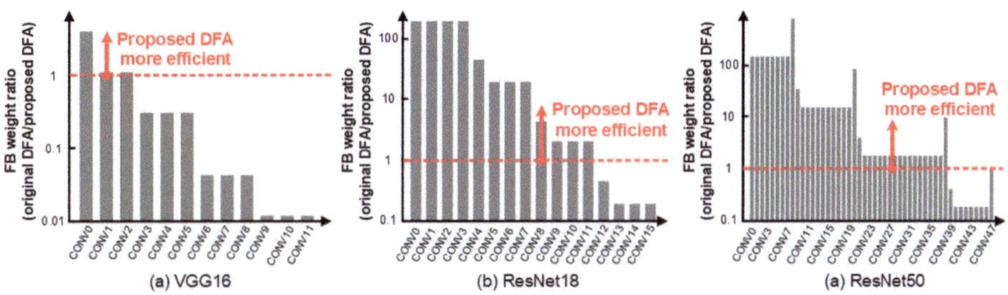

Fig. 4.8 Comparison of feedback weights according to two different DFA methods

from the direct error propagation path, the backward weight used in the last layer is initialized with the transposed form of its forward weight. Of course, the backward weight is determined at the start of the training and maintains its values until training is finished. In the LSTM and GRU training, there are multiple forward weights in the RNN cell, so we initialize the backward weight by using the average value of the multiple forward weights. The final performance improvement details that include the backward weight initialization are depicted in Fig. 4.7.

Effect of Backward Weight Sparsity

Figure 4.8 shows the training and test accuracy according to various backward weight sparsity. The effect of the sparsity becomes influential when the portion of

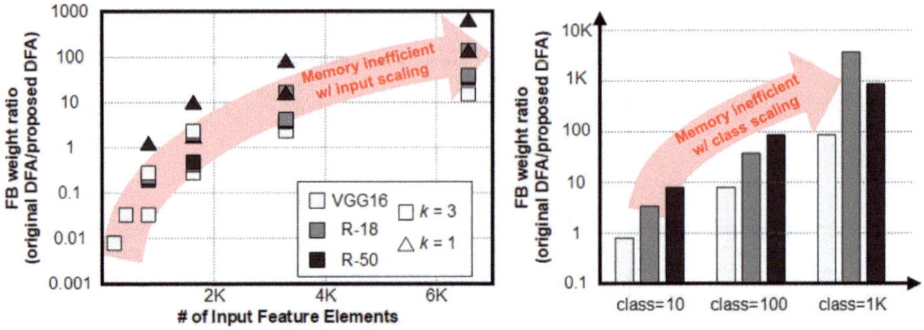

Fig. 4.9 DFA scaling issue with two different factors: input feature size (left), the number of classes (right)

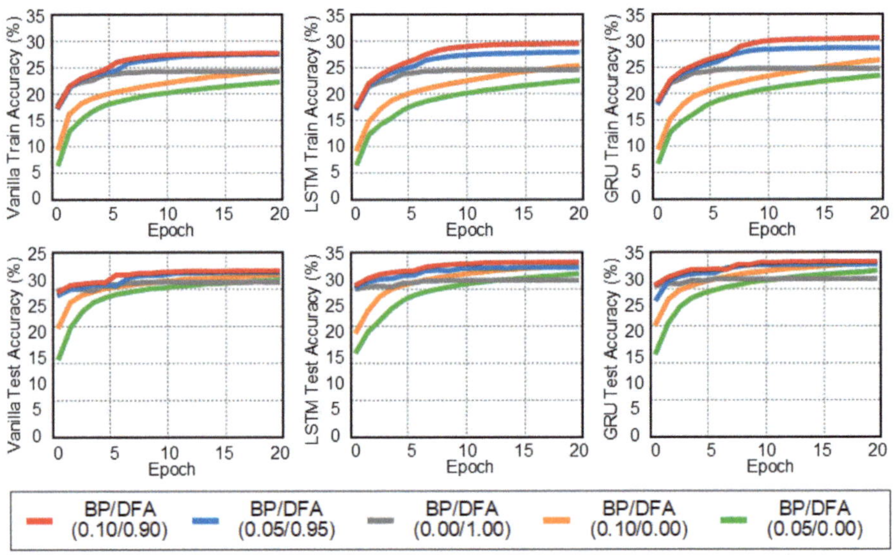

Fig. 4.10 RNN training and test accuracy curve with the proposed HDFA

BP is small. Additionally, vanilla RNN is vulnerable when the large sparsity appears in the backward weight. Another observation is that backward weight sparsity affects training accuracy but do not affect the test accuracy (Figs. 4.9, 4.10, 4.11, and 4.12).

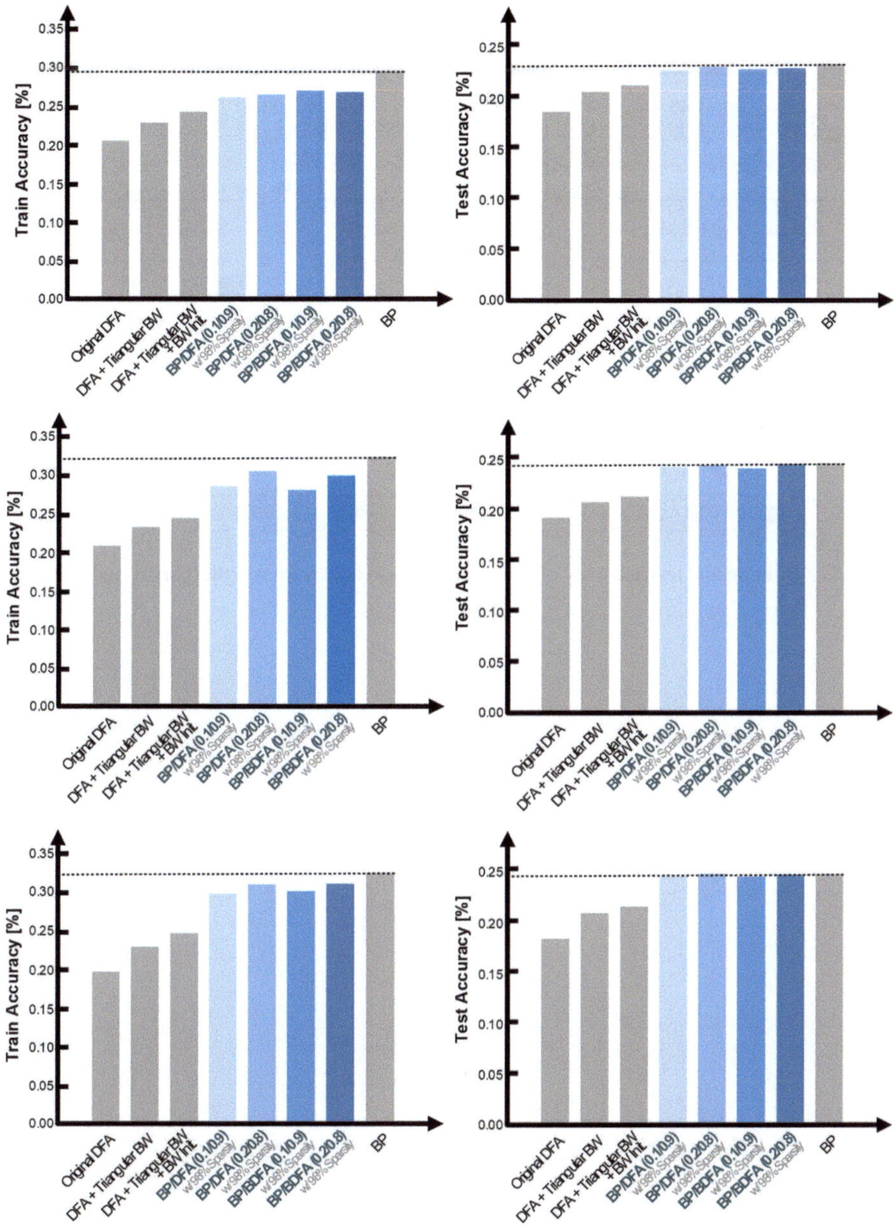

Fig. 4.11 Performance improvement details of DFA based RNN training

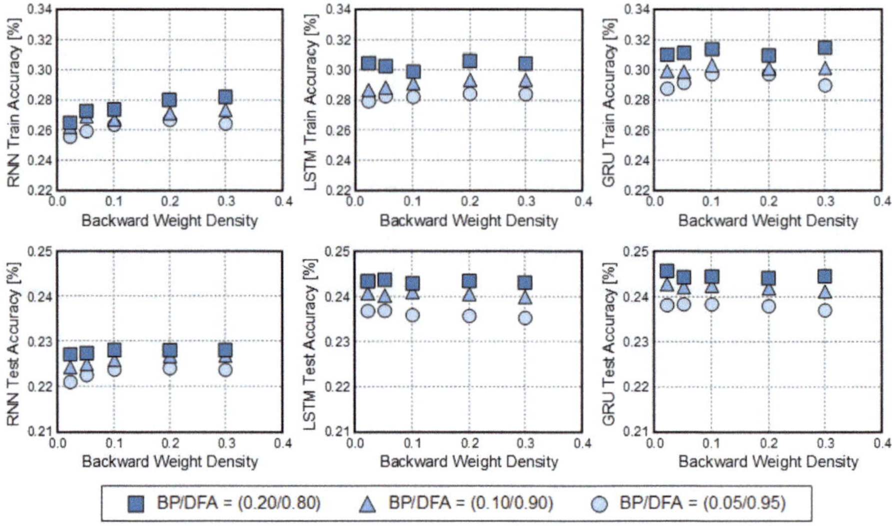

Fig. 4.12 Training and test accuracy of RNN according to backward weight sparsity and BP/DFA ratio

Chapter 5
DF-LNPU: A Pipelined Direct Feedback Alignment Based Deep Neural Network Learning Processor for Fast Online Learning

A deep neural network (DNN) learning processor, DF-LNPU, is proposed for fast online learning by utilizing direct feedback alignment (DFA). The proposed processor develops the pipelined DFA (PDFA) and shows 1.75 to 3.05× faster DNN learning speed compared with the previous processors. The processor consists of two different types of core. Thanks to the heterogeneous characteristics, the DF-LNPU shows higher area- and energy-efficiency than the homogeneous approach. Besides, inter-core and intra-core pipeline is constructed for high-speed online learning and reduces the overall processing time by 42.8–68.4%. At last, the direct error propagation core (DEPC) is proposed with the built-in pseudo-random number generator (PRNG). The DEPC adopts binarized DFA and organizes the adder-only computing units to maximize computation efficiency. The PRNG based backward weight generation reduces overall external memory access by 42.8%, and adder-only error computation improves the area- and energy-efficiency by 35.0% and 14.3%, respectively. The DF-LNPU is implemented in 65-nm CMOS technology and it can be operated from 0.7 to 1.1 V supply voltage with a maximum frequency of 200 MHz. The PDFA based computing enables 80.94 OP/KB throughput per memory bandwidth which is the best figure compared with the back-propagation dependent learning processors. The functionality of the DF-LNPU was successfully demonstrated on the verification system using MDNet based object tracking application.

This chapter is based on paper "DF-LNPU: A Pipelined Direct Feedback Alignment-Based Deep Neural Network Learning Processor for Fast Online Learning" (IEEE Journal of Solid-State Circuits, [34]).

© The Author(s), under exclusive license to Springer Nature Switzerland AG 2023
D. Han, H.-J. Yoo, *On-Chip Training NPU - Algorithm, Architecture and SoC Design*,
https://doi.org/10.1007/978-3-031-34237-0_5

5.1 Introduction

Artificial intelligence has been getting stronger and becomes familiar to human life with the dramatic development of the deep learning algorithm. Nowadays, deep neural network (DNN) has been conquering both image recognition [81] and natural language processing [122] challenges, and the related applications are getting larger and larger. With the huge usage of DNN applications, the development of the hardware becomes important to support the high-speed and energy-efficient DNN computing.

In the beginning, the DNN was too big to be computed in the mobile devices. Instead, a cloud server executed the DNN computations and the user could get the inference results. However, cloud-dependent DNN application faced the privacy problem because the users have the risk of sending their private data. To resolve this problem, the inference part of the DNN was moved to the edge devices. Since the edge devices have a small form factor, they have limited battery capacity. For this reason, DNN accelerators were developed to improve the energy-efficiency by using low-bit quantization and look-up table based computing [3, 4, 6]. Although the energy-efficient DNN inference engines solved the privacy issues that occurred during the server-edge communication, it only gave the inference results of the pre-defined object because the pre-trained model was based on a dataset collected by the server. On-chip learning can be a solution to those problems by tuning the network to user-specific tasks.

Figure 5.1 shows two typical examples of on-chip learning. The first intuitive example is transfer learning which modifies the last few fully connected layers (FCLs) to classify the new object defined by users. Besides, transfer learning can make up for the accuracy drop caused by the user-specific environment. The second application is the object tracking which combines the online learning concept to the object classification network. Even though the object classification networks [93, 94, 116] have shown good accuracy for steady objects, the performance can be degraded when they follow the abruptly deformed objects. For example, MDNet [41] as described in Fig. 5.2 tried to utilize the DNN online learning to resolve this problem, and it finally achieved the highest accuracy compared with the other conventional object tracking algorithms [124–126].

On-chip learning can provide new user-friendly applications with higher accuracy, but it gives new challenges to the DNN accelerators. First of all, the DNN accelerator should support not only inference but also training functionality. The second challenge is the high-bit-precision requirement. The low-bit quantization based DNN computing becomes difficult because the weight of the network can be changed at every moment. The last challenge is external memory management. DNN learning should handle multi-batch computing and repeat several iterations. Thus, it induces large data transactions of the intermediate feature maps and weights. It causes much more external memory access (EMA) compared with the DNN inference scenario. Since the data access from DRAM consumes 280 × more

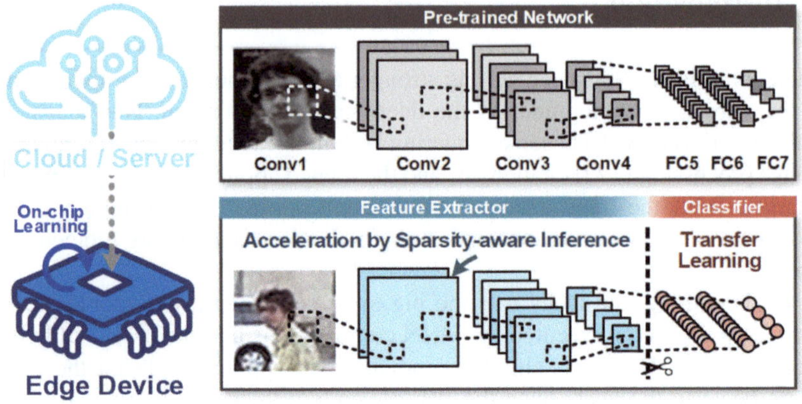

Fig. 5.1 On-chip learning examples in the edge devices 1: transfer learning for user-specific classification and domain adaptation

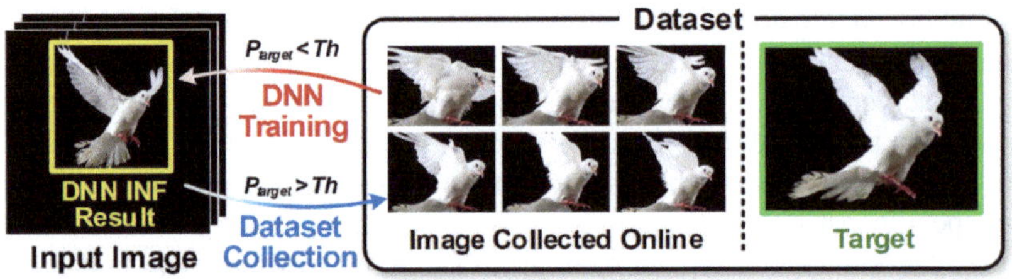

Fig. 5.2 On-chip learning examples in the edge devices 2: object tracking by combining object detection with DNN online learning [41]

energy compared with the multiply-and-accumulation (MAC) operations [127], the EMA reduction during DNN learning becomes the essential component.

The previous DNN learning processors [21, 123, 128] tried to beat those problems, but all of them have some limitations to achieve high-speed learning. The LNPU [21] suggested an input-sparsity-aware convolutional neural network (CNN) learning processor which achieved high throughput despite the small number of MAC units. Although it gave the opportunity of CNN training in mobile devices, it could suffer from EMA problem. Most of the CNNs are composed of feature extractor and classifier layers. The classifier is placed in the output region, and it is usually constructed with a few EMA dominant layers such as FCLs. We should note that most of the on-chip learning scenarios do not require scratch learning but do transfer learning, which performs the training of the last few layers. When the LNPU is applied to such an FCL training, the performance of the LNPU is degraded due to the EMA bottleneck. On the contrary, the authors in [123] focused only on FCL training. It adopted the run-length coding to compress the intermediate data required for DNN training and proposed a new algorithm, binary feedback alignment to minimize the EMA required for error propagation (EP) stage. However, 57.1% of the EMA still remains in the weight-gradient-update (WG) stage and shows poor energy-efficiency due to the low data reusability.

Another obstacle to high-speed learning is the backward locking (BL) problem. Conventional processors [14, 15, 21, 123, 128] accelerated the DNN training based on the back-propagation (BP). However, BP delays the next inference because it should wait until the BP of all layers is completed. The BP was taken for granted to train the network, but it now becomes the main reason for low-speed learning due to the BL. There were some trials [69–72, 129] to overcome this problem, but there was no approach to combine with the hardware solutions.

In this chapter, we proposed the new DNN training processor, DF-LNPU, with the two different types of learning core. The one is an input-sparsity-aware convolution core and the other is an FCL core. The FCL core can support not only back-propagation (BP) but also direct feedback alignment (DFA) algorithm to achieve high-speed learning. Additionally, the energy-efficient and high-speed on-chip learning processor is proposed with the following four key features: (1) a new training method, pipelined DFA (PDFA) for BL-free DNN training, (2) a heterogeneous core architecture for energy-efficient DNN training acceleration, (3) an inter-core and intra-core pipeline design, and (4) a pseudo-random number generator (PRNG) based direct error propagation core (DEPC) for both EMA and power optimization during the error propagation stage.

We organized the rest of the chapter as follows. In Sect. 5.2, conventional DNN learning algorithms are briefly introduced, and Sect. 5.3 introduces a new training method, PDFA. The overall architecture of the DF-LNPU and the key building blocks are explained in Sects. 5.4 and 5.5, respectively. Section 5.6 shows the chip implementation results and the chapter will be concluded in Sect. 5.7.

5.2 Related Algorithms

Before explaining the details of the DF-LNPU, the basic notations and summary of the DNN learning algorithms will be described. First of all, the three main stages of DNN learning will be explained. Then, the difference of the BP and DFA based learning algorithms will be discussed with the BL problem.

5.2.1 Back-Propagation and Backward Locking Problem

Generally, the DNN adopts BP based learning algorithm, and the algorithm consists of three different learning stages (Fig. 5.3, left). The first stage is feed-forward (FF) where the network gets the classification results by going through the multiple layers. Each layer multiplies an input activation $X_{(i,t)}$ and a weight $W_{(i+1,t)}$ to generate the output activation $X_{(i+1,t)}$. After multiplication, the output activation experiences an activation function such as ReLU [95]. After the FF stage, the final

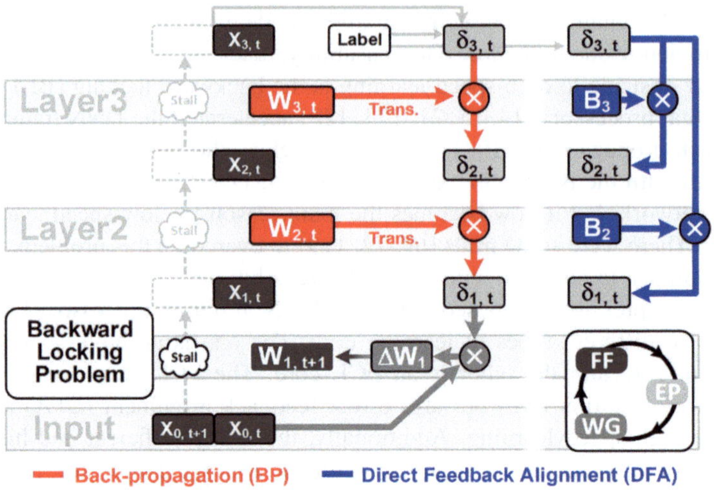

Fig. 5.3 Summary of back-propagation and direct feedback alignment [71, 72]

output $X_{(3,t)}$ is used as the input of softmax function, and it generates the error $\delta_{(3,t)}$ by comparing with the pre-defined label. The error should be computed to evaluate the gradient of the weight, so it starts the error propagation (EP) stage to deliver the error of the last layer to intermediate layers. In the BP based learning, the error is propagated gradually from the last layer to the first layer. During the BP, weight $W_{(i+2,t)}$ used in the FF stage should be transposed and $W_{(i+2,t)^T}$ is multiplied with the $\delta_{(i+2,t)}$ to make $\delta_{(i+1,t)}$. When $\delta_{(i+1,t)}$ is generated, weight-gradient-update (WG) stage can be started. In this stage, the network aggregates $X_{(i,t)}$ and $\delta_{(i+1,t)}$ of all batch data and makes the gradient $\Delta W_{(i+1,t)}$ (gradient generation, GG). The $\Delta W_{(i+1,t)}$ will be used to make a new weight $W_{(i+1,t+1)}$ by using gradient optimizer such as momentum (weight update, WU) [100]. These three learning stages, FF, EP, and WG, are repeated several iterations to complete the DNN training.

The problem of BP is the sequential computation. In the BP based learning scenarios, three different learning stages cannot be computed in parallel. The FF of the next iteration can be started just after WG is finished, but the error of the first layer can be determined after the EP of the later layers. In summary, the new FF can be started only when the EP of the entire layer is finished. This sequential propagation problem is called "backward locking" (BL).

5.2.2 Direct Feedback Alignment

Direct feedback alignment (DFA [71, 72]) is a typical attempt to solve the BL problem. Even though the BP propagates errors from the last layer to the prior layer in a sequential manner, the DFA propagates errors directly from the last layer

(Fig. 5.3, right). Instead, it generates the new backward connections between the last layer and the intermediate layers. In this process, the DFA requires a new weight, B, which is used only in the EP stage. The backward weight (BW) B can be the random value but maintains its value once the network is determined. Except for the EP, the other learning stages are the same as BP.

5.3 Pipelined Direct Feedback Alignment

The DF-LNPU supports not only BP but also DFA based high-speed learning. The processor utilizes a new learning method, pipeline DFA (PDFA), which is the new pipeline structure enabling parallel computing of the three learning stages based on the DFA. Let us assume that the FF of the first iteration is finished and the errors of the last layer are computed. In the DFA algorithm, the error of the first layer can be propagated directly from the last layer. After the EP of the 1st layer, the WG stage of the corresponding layer can be started. In the meantime, the EP of the 2nd layer can be computed independently. When WG of the 1st layer and EP of the 2nd layer is done, FF of the 1st layer can be started. As a result, three learning stages, FF, EP, and WG are in charge of the different layers and they can be computed simultaneously as shown in Fig. 5.4. The PDFA overcomes the problem of the slow learning speed caused by BL but the required computation is the same as the conventional DFA algorithm. The difference of the PDFA is the computation order. The conventional DFA tries to compute errors of multiple layers simultaneously, but the PDFA performs the parallel processing of the three learning stages.

During the on-chip learning scenarios, the most DNN training occurs at the later layers which have the characteristics of the FCLs. We verified that the PDFA algorithm does not have an accuracy drop in those scenarios [72]. The PDFA can rather show better performance compared with the BP based DNN training. The accuracy of the PDFA can be dropped only in the scratch learning scenario because the PDFA shows poor performance if it is applied to the prior convolutional layers (CLs). To overcome this problem, the DF-LNPU supports both BP and DFA based

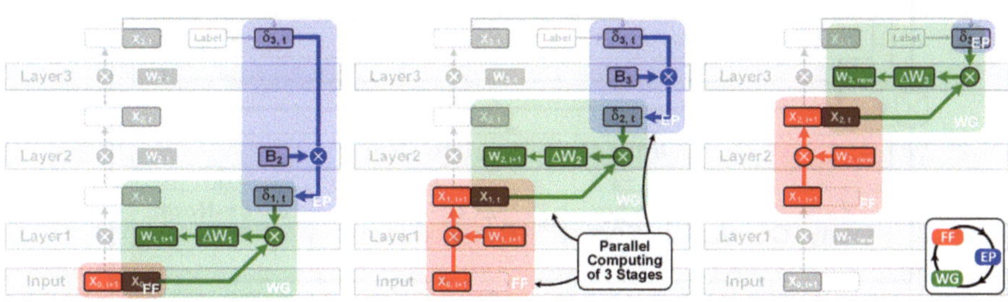

Fig. 5.4 Proposed DNN learning method: PDFA

learning methods to the last FCLs and utilizes only the BP algorithm to the prior CLs. The DF-LNPU can select the learning method of the FCLs according to the on-chip learning scenarios.

5.4 Overall Architecture

Figure 5.5 describes the overall architecture of the DF-LNPU. From a macro view, the DF-LNPU can be divided into two cores: (1) two zero-skip convolution cores (ZCCs) and (2) a PDFA based learning core (LC). ZCCs are in charge of inference and training of the CLs and the LC computes the FCLs by using the PDFA methodology.

When the PDFA based learning is performed, the EP core computes loss of the FF results and determines the error, $\delta_{(3,t)}$. The errors of the intermediate layers are generated by the DFA algorithm and they are transmitted to the GG core. The GG core makes the weight gradients of the corresponding neuron connections. The GG core collects all batch data such as activations and errors in a transposed manner. The final multiplication results in the GG core are moved to the weight update unit (WUU). The role of WUU is receiving the weight from the external DRAM and computes the momentum optimizer to make a new weight. The double buffered weight memory (WMEM) is used as the temporal buffer to communicate with the

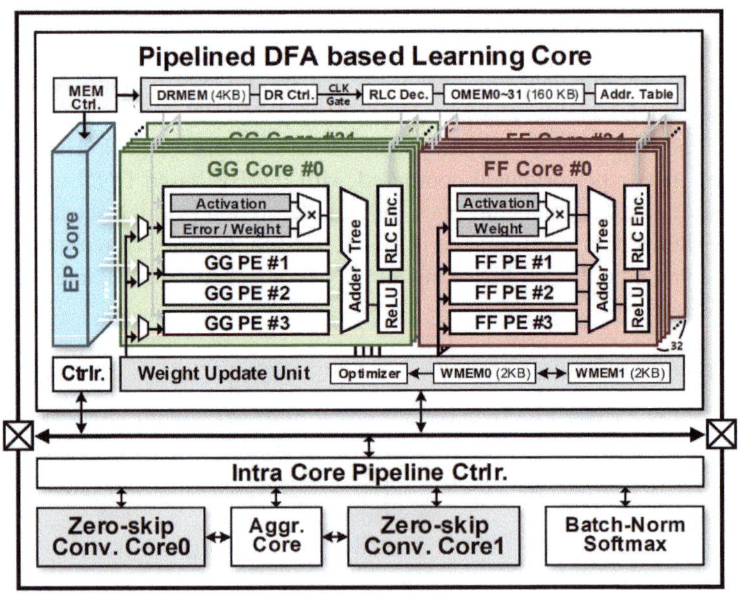

Fig. 5.5 Overall architecture of the DF-LNPU

external DRAM. The new weight generated by WUU is transferred to the FF core, and the FF core executes the inference of the next iteration.

All intermediate activations and errors are stored in the output memory (OMEM). The OMEM consists of 32-bank SRAM and the total size is 160 KB. The run-length coding (RLC [123]) decoder is placed near the OMEM, so both the FF and GG cores receive the decoded activations. Before the PDFA based learning, the input of the FCLs should be fetched into OMEM. This input data can be directly transferred from the aggregation core or loaded from the external DRAM. In the latter case, DF-LNPU fetches data from the pre-collected dataset to avoid the overfitting problem. In the online learning scenario, the DF-LNPU occasionally stores the input data to the DRAM and utilizes them to the later DNN training. The number of datasets cumulated in the DRAM and the data collection methodology highly depend on a target application.

The LC supports the dropout regularization and utilizes clock-gating to minimize power consumption. The clock-gating methodology is the same as [123]. A dropout controller receives the dropout indicators from the dropout memory (DRMEM). The dropout indicators can be modified when the derivatives of the ReLU activation function are reflected. This modification enables the EP of the ReLU activation function. The dropout controller prevents the power consumption when the output value is predicted as zero. Thanks to clock-gating, overall 23.1% of the power consumption is reduced during the inference-only scenario. On the other hand, 16.4% of the power consumption can be saved during online learning.

The DF-LNPU adopts both inter-core and intra-core pipeline designs. Inter-core PC has the role of managing the schedule of CL computing and PDFA based FC computing. The pipeline controller (PC) placed in the LC manages the schedule of the three different learning cores. The details of the PC will be explained in Sect. 5.5.2.

5.5 Proposed Processor: DF-LNPU

In this section, the key features and building blocks of the processor will be introduced with the implementation detail. Firstly, heterogeneous core architecture will be described with the inter-core and intra-core level pipeline design. The energy-efficient EP core with built-in PRNG will follow.

5.5.1 *Heterogeneous Learning Core Architecture*

The DF-LNPU handles two different kinds of layers with the heterogeneous core architecture. The ZCC focused on throughput improvement with the limited number of MACs and maximizes efficiency. In contrast, the LC adopts the PDFA based learning method and improves both data reusability and throughput per bandwidth.

Zero-Skip Based Convolution Layer Optimized Core

The earlier CNN accelerators [1, 3] focused on the characteristics of the convolution operation and maximize the data reusability. Also, the precision requirement of the CNN accelerator is not strict for inference-only cases, so they utilized quantization and look-up table based computing for higher energy-efficiency. However, the work [59] proves that the precision requirement for DNN training is more strict compared with the inference. Our convolution core finally adopts FP 8-bit precision for MAC operation based on an experiment done by Wang et al. [56]. Instead, the low energy-efficiency caused by high bit-precision is compensated by input-sparsity-aware core design.

Generally, most of the CNNs utilize the ReLU activation functions because of simple and strong performance. The ReLU activation induces many zeros to input activations. Since the multiplication with the zero values does not affect the final output result, the processor can skip the multiplication with the zero values. The DF-LNPU adopts zero-skipping core design proposed in the LNPU. The only difference is the adopted precision. Although the LNPU supports both FP 8-bit and 16-bit in a single MAC unit, the DF-LNPU is implemented with only FP 8-bit MAC units because of the different learning scenarios. Both the throughput and energy-efficiency are influenced by the input sparsity ratio and the detailed specification is described in Figs. 5.19 and 5.20. Thanks to the zero-skipping based convolution acceleration, it shows up to 3.34 $\times$ throughput improvement and enables high-speed computation of CLs.

PDFA Based Fully Connected Layer Learning Core

The core architecture of the LC focused on memory bandwidth optimization. The previous FC optimized inference accelerator solved the EMA problem by using the low-bit quantization [4, 6]. In the DF-LNPU, it cannot use the low-bit precision due to the on-chip learning functionality but use the input sparsity aware activation encoding method introduced in [1, 123]. The activation encoding minimizes the EMA caused by the activation transaction and the SRAM footprint used as the temporary cache of the computation. However, the training of the FCLs still suffers from the large memory access caused by the weight data transaction.

Figure 5.6 describes the general computation flow of the FC inference scenario. Since each connection between an input and an output neuron has a different weight, the weight is used only once during the computation. The FC core shows low energy-efficiency because of the low weight reusability. In the FCLs, low weight reusability is a common characteristic even in the other two learning stages, EP and GG. Previously, the authors in [123] suggested homogeneous core architecture to accelerate the three different learning stages. Thus, the size of the single core becomes huge to support the different data-path requirements of the three learning stages (Fig. 5.7).

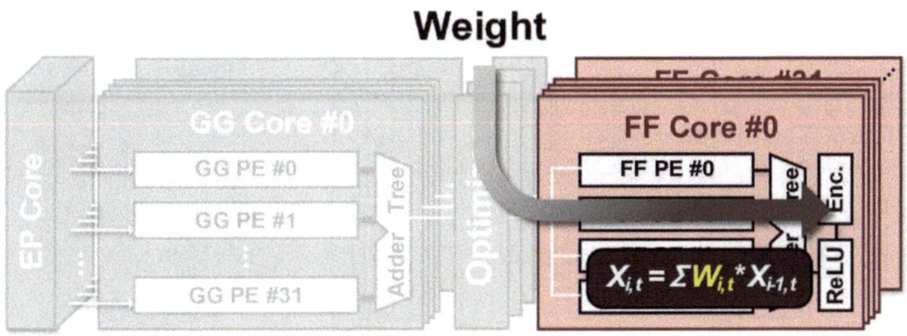

Fig. 5.6 Data-path of the weight during the feed-forward stage

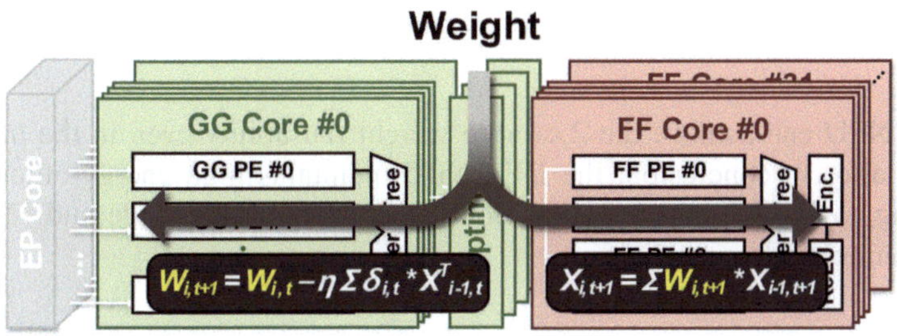

Fig. 5.7 Data-path of the weight during the online learning

The LC of the DF-LNPU shows the opposite strategy due to the PDFA. The PDFA enables parallel processing of the three learning stages, so a single core does not need to cover the various data-path required for training. For this reason, the GG core requires a transpose decoder to compute the gradient and the computation results should be transferred to the gradient optimizer. The EP core can be simplified with the binary DFA [72]. As a result, the composition of the FF, EP, and GG cores shows a big difference because of the different computation characteristics. The final area is described in Fig. 5.8.

The energy-efficiency improvement of the DF-LNPU is induced by not only the core optimization but also the larger data reusability. In the PDFA based learning scenario, the weight updated by GG core is used as the FF of the next iteration. Thus, the weight can be reused for both GG and FF core. The DF-LNPU shows 2× more computational reuses compared with the conventional processors thanks to the PDFA based parallel processing. Consequently, the heterogeneous core architecture can optimize a core and data-path to each learning stage, so it improves the area- and energy-efficiency by 84.2% and 58.6%, respectively (Fig. 5.15).

The DF-LNPU enables higher energy-efficiency by using the PDFA based parallel training, but it shows low efficiency during the inference-only scenarios. To resolve this problem, the DF-LNPU designed the GG core to support both GG and FF stages. Therefore, the GG core can be utilized as WG computation during the PDFA scenario, but it can compute the FF stage even in the inference-only scenario. Dual-mode GG core enables 2× more batch size during the inference and

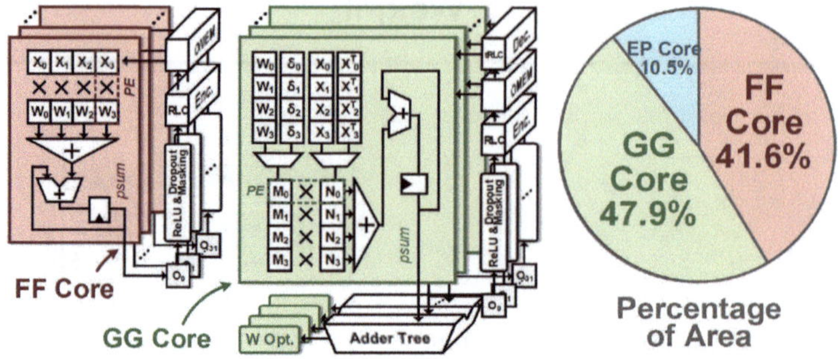

Fig. 5.8 Details of FF and GG core architecture and area occupation

the DF-LNPU can achieve the 2× more weight reusability even in the inference-only scenario. In conclusion, the DF-LNPU maintains high energy-efficiency by supporting two different modes: (1) double-batch inference mode and (2) single-batch training mode.

5.5.2 Inter-core and Intra-core Pipeline Design

The DF-LNPU adopts a heterogeneous core architecture and the scheduling of the cores is a key component of the throughput improvement. For this reason, both inter-core and intra-core PC are implemented to improve the on-chip learning throughput. As described in Fig. 5.5, an inter-core PC is placed between the ZCC and an LC, and it manages not only a pipeline schedule but also the data transaction between ZCC and LC. Intra-core PC is implemented inside the LC, and it builds a fine-grained pipeline with three learning stages.

Inter-core Pipeline Design

As illustrated before, the computation of CL is handled by the two ZCCs, but the LC has charge of computing in FCL. The workloads allocated for the ZCC and the LC can be different due to three factors. The first factor is the network configuration. There are various configurations of the CNN [124–126], so the ratio of CLs and FCLs varies widely. Second, the batch size and the number of iterations can be varied by the on-chip learning scenario. The last factor is input sparsity. The DF-LNPU adopts input-zero-skipping architecture for the higher throughput. Therefore, the performance can be varied if the zero ratios of the input activation are changed.

Although there are many factors which vary the workload between ZCCs and LC, the computation time of CLs is usually much larger than FCLs. For this reason, the inter-core PC takes a simple scheduling method as giving priority to the computation of CLs. The DF-LNPU first executes the CLs by using ZCCs. Since the LC needs the inference results of ZCCs, the LC can compute the FCLs after ZCCs complete their workloads. However, inter-core PC starts the ZCCs first to compute the next iteration even if the LC is ready to be executed. It is possible because the weights of the CLs do not need to be updated. Instead, the PC evaluates the EMA status and it gives the run signal to the LC when there is no data transaction between DRAM and ZCCs. The ZCC rarely requires the EMA because the weight and activation loaded from the DRAM are reused multiple times to complete the output calculation. Thus, inter-core PC utilizes the remained memory bandwidth to compute the FCLs, and it finally enables pipeline design between the ZCC and LC. Although the computation portion of FCLs is usually smaller than CLs, the processing time of the FCLs can be reversed depending on the size of batch or iterations. In this case, inter-core PC stalls the ZCCs until the LC completes the computation of FCLs.

Thanks to the inter-core PC, the DF-LNPU can improve the throughput by utilizing remained external DRAM bandwidth during the CLs. The inter-core pipeline does not affect the processing time of inference, but the computation time of the multi-iteration online learning can be reduced due to the parallel processing of the CLs (Fig. 5.9).

Intra-core Pipeline Design

Inter-core PC is in charge of scheduling between ZCCs and LC. In contrast, intra-core PC is placed inside the LC and manages the pipeline design of the three different learning stages. As illustrated before, BP suffers from the BL problem. Hence, the overall processing time of the FCLs can be significantly increased as

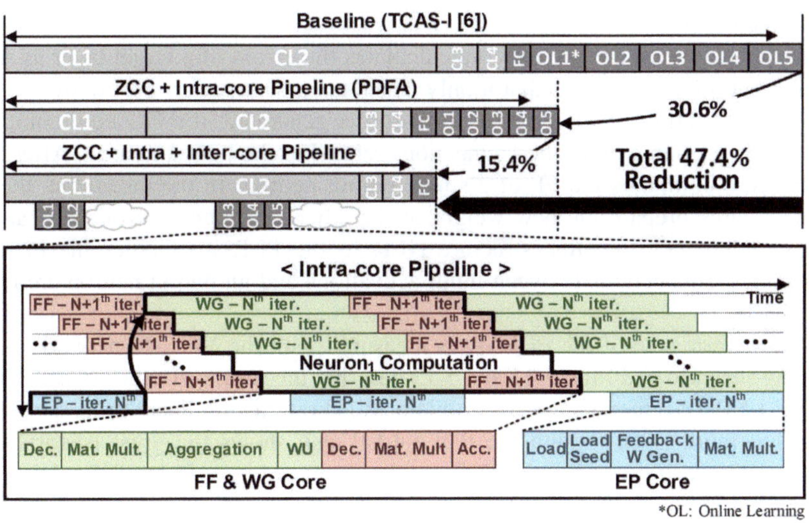

Fig. 5.9 Inter-core and intra-core pipelined design (example: VGG-M)

the number of iterations becomes large. The PDFA adopted in the DF-LNPU can improve the speed by building the pipeline of the three learning stages.

The intuitive pipeline design of the PDFA was explained in Sect. 5.3, and we call it the layer level pipeline (LLP). When the EP core computes the errors of the certain layer, the GG core and optimizer units generate the new weights of the prior layer. After the GG core generates the new weight, the FF core can start the computation of the next iteration. Consequently, three different cores in the LC compute the three different layers simultaneously in the LLP.

However, the LLP has two potential problems which disturb the high-speed and energy-efficient computing. First, the gradient generated by the GG core should be stored in the temporary buffer inside the LC. Since the size of weights required for FCLs is much larger than the activations, the LLP induces the large size of the temporary buffer, which can cause a large SRAM footprint and low energy-efficiency. The other problem is the low core utilization. Since the PDFA with LLP makes each core manage the distinct layer, LLP cannot be constructed if the number of FCLs is small. Moreover, the FF core can be executed only after the computation of the EP and WG stages is finished. Until the WG stage completes the new weight generation of the first layer, there is no task allocated in the FF core. As a result, the LC shows only 58.5% core utilization during the PDFA-LLP based VGG-M online learning scenario.

To solve these problems, we should remember the principle of FCL computations. A connection between an input and an output activation has its own weight which will be multiplied with the input activation. The value of the corresponding weight is independent of other connections. In other words, if the weights of some connections are ready for new FF, the DF-LNPU can start the FF computation even though some connections of a layer are not ready. Thanks to this property, the PDFA is evolved to advanced pipeline design, neuron level pipeline (NLP), and partial-sum level pipeline (PLP).

In the NLP and PLP, the EP core calculates the errors of a target layer as same as LLP, but the difference is that it only generates the partial errors required for the next FF computing. If the error of a neuron is generated, the GG core computes the weight gradients of all the connections related to the output neuron. After that, the FF core computes the FF of a single output neuron. In the meantime, the EP and GG cores prepare the new weight of the adjacent output neuron. We call the corresponding new pipeline structure NLP. In the PLP, it executes the FF core after partial connections between input activations and an output neuron are ready (Fig. 5.10).

The performance of the PDFA was measured with the VGG-M network online learning scenario and Figs. 5.11 and 5.12 describe the results. Thanks to the PDFA, the DF-LNPU overcomes the BL problem and shows a significant processing time reduction ratio. Even though the PDFA-LLP also shows the great performance, the required cache to store the WG results is quite large. Both PDFA-NLP and PDFA-PLP show similar FCL learning speed, but the minimum required WMEM is much smaller in the PLP. In addition, conventional BP based training requires large OMEM to store all errors generated during the EP stages. The PDFA needs

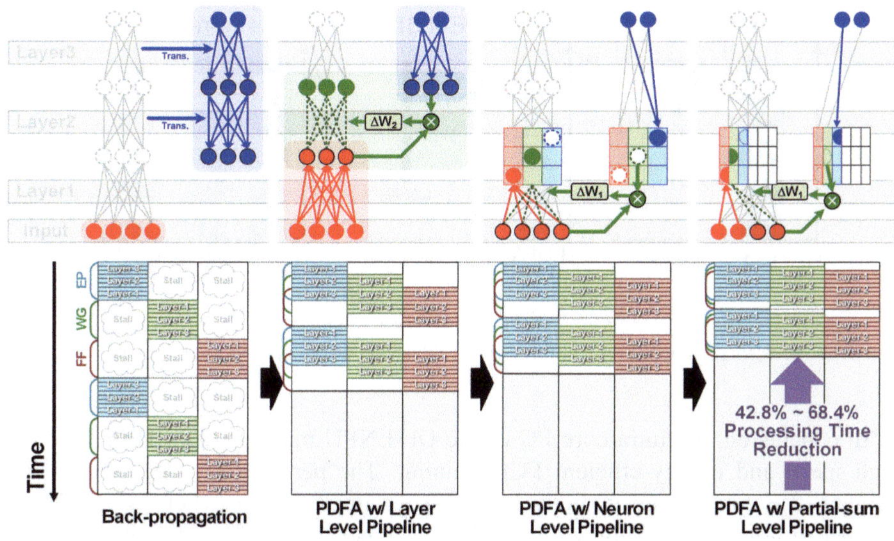

Fig. 5.10 Advanced intra-core pipeline design with PDFA

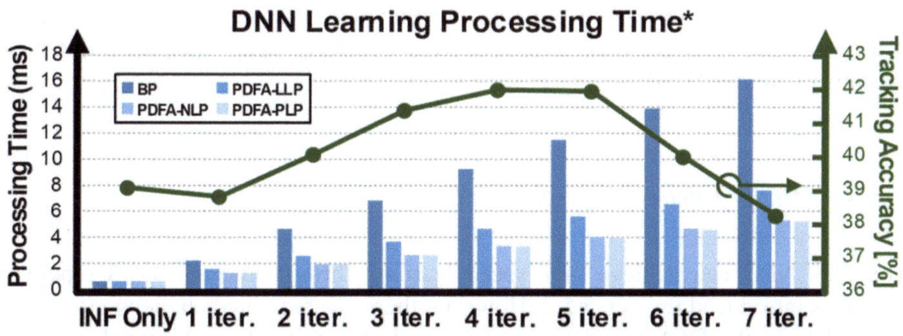

	INF Only	1 iter.	2 iter.	3 iter.	4 iter.	5 iter.	6 iter.	7 iter.
BP	0.6624	2.305	4.609	6.914	9.219	11.524	13.828	16.133
PDFA-PLP	0.6624	1.319	1.982	2.644	3.307	3.969	4.632	5.294
Speed	-	× 1.75 ▲	× 2.33 ▲	× 2.61 ▲	× 2.79 ▲	× 2.90 ▲	× 2.99 ▲	× 3.05 ▲

Fig. 5.11 Performance of the advanced intra-core pipeline according to iteration number

a small buffer because it only requires to store the error that appeared in the last layer. The error required for WG is directly generated by DEPC and the detailed operation will be described in Sect. 5.5.3. Finally, the DF-LNPU can reduce 33.3% of the required OMEM by adopting PDFA. The DF-LNPU with PDFA-PLP also shows higher core utilization compared with the PDFA-LLP. The utilization can be varied when the number of neurons in the last layer is small. It shows 82.7% and 92.6% core utilization when the number of neurons is two and ten, respectively.

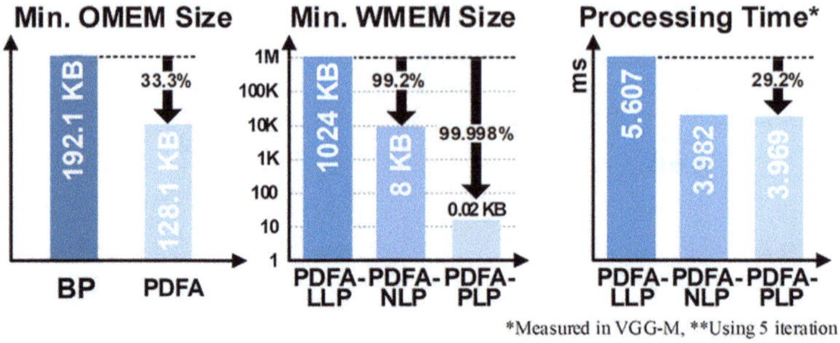

Fig. 5.12 Performance improvement through the advanced intra-core pipeline

In conclusion, the intra-core PC of the DF-LNPU manages the PDFA-PLP for high-speed and energy-efficient FCL training. The performance depends on the number of iterations and it shows higher speed improvement when using a large number of iterations. It shows 1.75 to 3.05× higher training speed compared with the BP based training. In addition, the DF-LNPU reduces the required buffer size of OMEM by 33% and removing the overhead of weight buffer for high-speed learning. When the intra-core pipeline is combined with the inter-core pipeline, it achieves 47.7% processing time reduction compared with the previous DNN learning processor [123].

5.5.3 Direct Error Propagation Core

The DF-LNPU can achieve high-speed learning through the advanced pipeline design as introduced in Sect. 5.5.2. However, there is a remained challenge to apply the PDFA. In the DFA based learning method, it requires the additional weights in the EP stage. In this chapter, we define the weight used during the FF stage as the forward weight (FW) and the weight used during the EP stage as the backward weight (BW). In the PDFA based learning, the FW affects the inference results and it should be updated after the WG stage. On the other hand, the BW affects the value of errors and the BW should be maintained even after the WG stage. The value of BW can be the collection of random values, but it does not mean the value can be changed at the next iteration. The BW can be loaded from the DRAM, but it induces extra EMA addition to the FW. Therefore, the PDFA requires higher external memory bandwidth because both FW and BW should be loaded simultaneously (Fig. 5.13).

To minimize the overhead due to BW, the DF-LNPU adopts the binary DFA concept introduced in [72]. In the binary DFA, the BW can be represented with the 1-bit data and it indicates only the sign value. The DF-LNPU implements the DEPC which is an optimized core design to apply the binary DFA. The DEPC calculates the errors which will be used for the WG core. The error appeared in the last layer

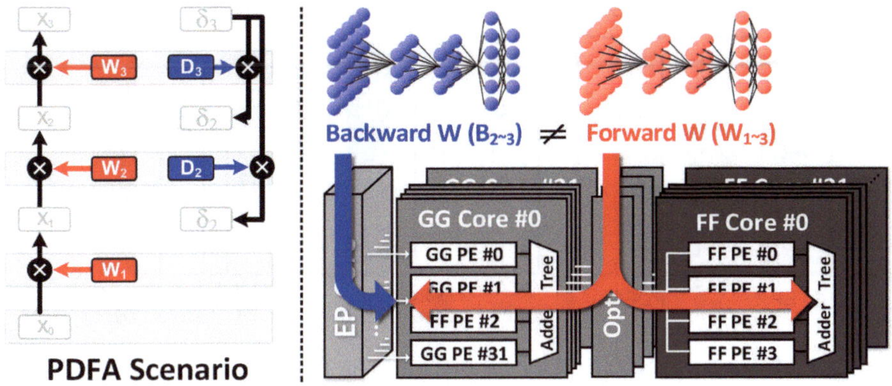

Fig. 5.13 Heterogeneity of forward and backward weight

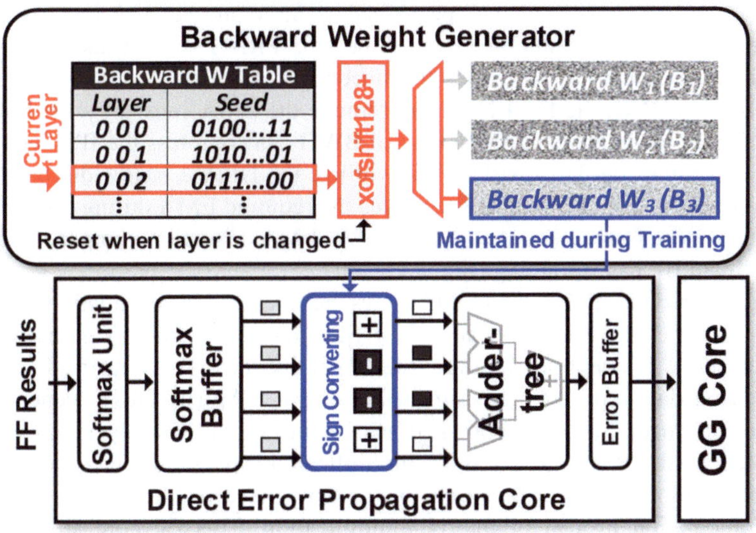

Fig. 5.14 PRNG based backward weight generator and direct error propagation core

and the BW is used to generate the errors of the intermediate layers. At this moment, the BW is loaded from the adjacent backward weight generator (BWG). The BWG includes the built-in PRNG which makes the pre-determined BW by referring to the BW seed table (Fig. 5.14).

The operation scenario of the BWG is described as follows. At first, the BWG receives the layer number that the EP core wants to calculate. Then, the corresponding random number seed will be loaded from the BW seed table. The seed is used as the input of the xorshift128+ unit [130]. After receiving the random number seed, the xorshift128+ unit generates random number sequences. When the

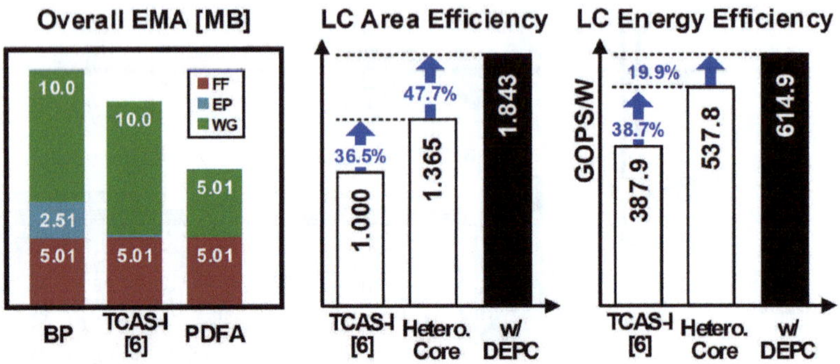

Fig. 5.15 EMA reduction and efficiency improvement of the DF-LNPU

EP of the certain layer is finished, the DF-LNPU resets the PRNG and transfers the other random seed. Once the operational scenario and network are determined, the seeds stored in the table are fixed. Since the PRNG is reset and receives the constant seed, the BWG generates the unchanged BW for all iterations. Thanks to the built-in RNG, the DF-LNPU can remove the EMA caused by the BW with the minimum hardware cost. PRNG based BWG can compress the BW with a few bytes and enables the DF-LNPU to fully utilize the DRAM bandwidth on the FW data transaction. The BWG totally removes the required EMA during the EP stage, and the proposed processor shows the 42.8% EMA reduction compared with the conventional BP based training (Fig. 5.15, left).

As described in Fig. 5.8, the DEPC shows the smallest area occupation compared with the other cores. That is because the DEPC adopts the adder-tree based matrix multiplication. As mentioned before, the BWG decodes BW by using PRNG and the decoded data represents the sign value of the BW. The DEPC generates the intermediate error by using both softmax results of the last layer and the BW generated by the BWG. There is no multiplication operation, but it converts the sign information of the error based on the BW. As a result, it performs adder-only computation instead of matrix multiplication. This EP core simplification maximizes the efficiency of the heterogeneous core architecture. It shows a 35.0% higher area-efficiency and 14.3% higher energy-efficiency.

5.6 Measurement Results

Figure 5.16 shows the chip implementation results and Fig. 5.18 shows the performance summary (Fig. 5.17). The DF-LNPU is fabricated with the 65-nm 1P8M CMOS technology with a 4 $\times$ 4 mm2 area occupation. Each ZCC occupies 2.81 mm^2 area, while the LC occupies 5.36 mm^2 (Fig. 5.18). The size of ZCC is quite reduced compared with the conventional LNPU, and the difference is induced by the different number representations.

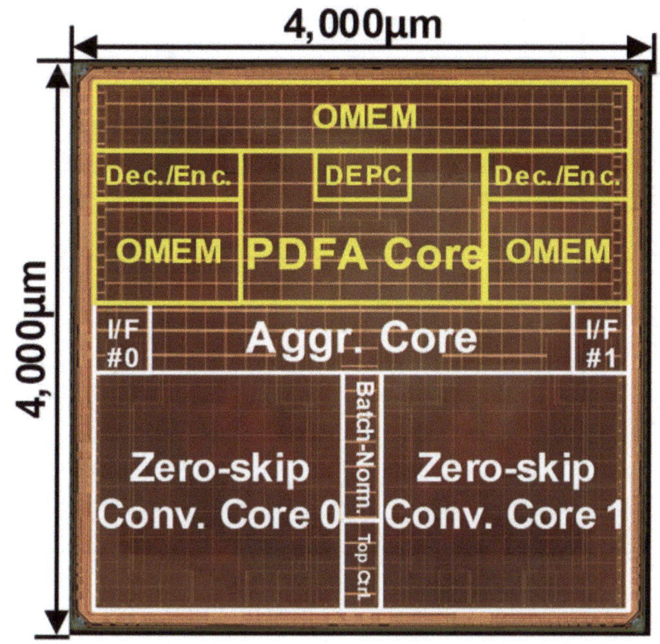

1) @ Sparsity: 0%, 2) @ Sparsity: 90%

Fig. 5.16 Chip photograph of DF-LNPU

The ZCCs operate from 50 to 200 MHz frequency and the corresponding energy-efficiency can be varied in accordance with the input sparsity ratio. The energy-efficiency is measured as 3.7 TOPS/W when there is no zero on the input activations. If the sparsity is increased up to 90%, the energy-efficiency can be improved until 27.1 TOPS/W. The detailed throughput and energy-efficiency improvement results depending on various sparsity are illustrated in Figs. 5.19 and 5.20. Each ZCC includes the 84.5 KB SRAM to store both activations and weights.

The LC can operate from 25 to 200 MHz frequency with 0.7-to-1.1 V supply voltage. The detail of voltage–frequency scaling graph is depicted in Fig. 5.17. Although ZCCs adopt the floating-point arithmetic units, the LC adopts fixed-point representation. The LC uses 16-bit representations for both activation and weight, but 3-bit is used to indicate the zero length in the activation representation. The LC includes a larger buffer to store all intermediate activations but a smaller number of MAC units because the number of MACs is determined by the given memory bandwidth. The LC shows 155.2 GOPS throughput and 1.1 TOPS/W energy-efficiency at the maximum core utilization.

The performance of the DF-LNPU was measured with the object tracking network, MDNet [41], and the detailed performance benchmark is described in Fig. 5.21. In this scenario, the ZCCs accelerated four CLs and the LC was in charge

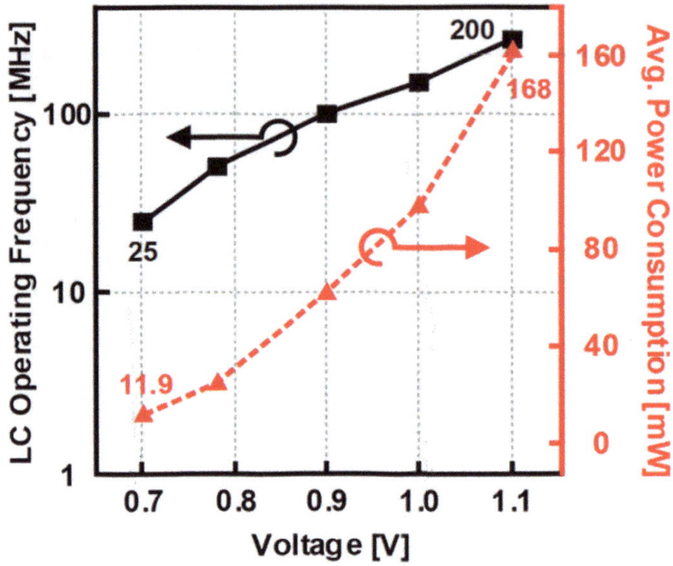

Fig. 5.17 Dynamic voltage–frequency scaling results of DF-LNPU

	Specification	
Technology	65 nm Logic CMOS	
Die Area	4.00 mm × 4.00 mm (16.00 mm^2)	
Online Learning Core (LC)		
Die Area	3.57 mm × 1.50 mm (5.36 mm^2)	
SRAM	168 KB	
Precision (A/W)	FXP 13b / FXP 16b	
Operating Condition	**200MHz @1.1V**	**25MHz @0.7V**
MAX. Throughput	155.2 GOPS	19.4 GOPS
Power Consumption	252.4 mW	17.9 mW
Energy Efficiency	614.9 GOPS/W	1083.8 GOPS/W
Zero-skip Convolution Core (ZCC)		
Die Area	1.70 mm × 1.66 mm × 2 core (5.64 mm^2)	
SRAM	84.5 KB / core	
Precision (A/W)	FP 8b / FP 8b	
Operating Condition	**200MHz @1.1V**	**50MHz @0.78V**
MAX. Throughput	300~1002.1 GOPS	75~250.5 GOPS
Power Consumption	171.6[1)]~78.5[2)] mW	20.2[1)]~9.3[2)] mW
Energy Efficiency	1,748[1)]~12,766[2)] GOPS/W	3,722[1)]~27,072[2)] GOPS/W

Fig. 5.18 Performance summary of DF-LNPU

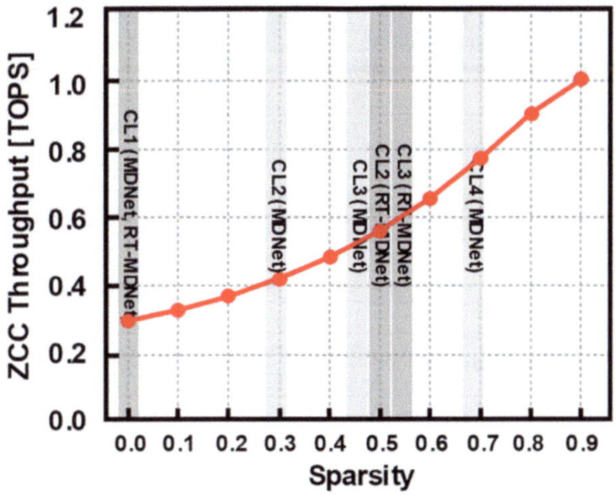

Fig. 5.19 Throughput of ZCC according to input sparsity

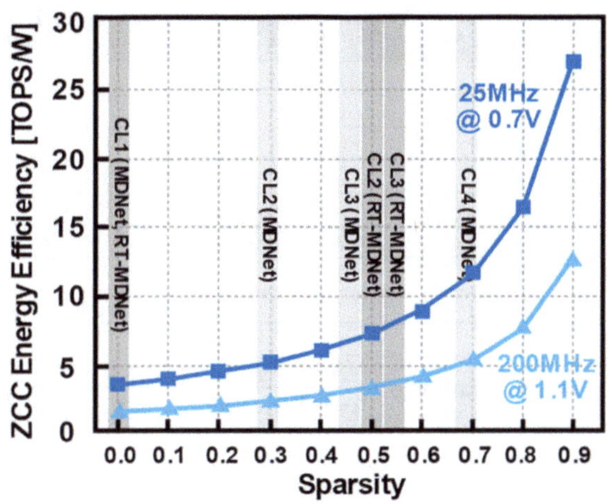

Fig. 5.20 Energy-efficiency of ZCC according to input sparsity

of online learning of three FCLs. The throughput of the ZCCs is varied due to the sparsity difference among the layers. During the online learning, the DF-LNPU uses 32 batches to two- or five-iteration learning. Thanks to zero-skipping convolution core and PDFA-PLP based FCL training, the processing time of the benchmark is reduced by 47.4% and it achieves an average of 218 mW power consumption. Additionally, we added recent object tracking networks, RT-MDNet [131] and Siammask [132]. The DF-LNPU shows average 17.5 frames per second (FPS) and 12.7 FPS during RT-MDNet and Siammask based object tracking. In the Siammask based tracking, it does not utilize online learning but shows lower framerate and energy-efficiency because of the huge computation amount. The DF-LNPU shows an average of 3.00 TOPS/W energy-efficiency during the inference of Siammask. We also measured the performance of the DF-LNPU in the autonomous driving application. A typical autonomous driving network, DroNet [133] used ResNet34

MDNet	Kernel Size	Input Size	In/Out Channels	Params.	MAC / batch	Effective GOPS	Processing Time [ms]
Conv1	7×7	107×107	3/96	13.8 KB	36.7 M	300.0	7.83
Conv2	5×5	25×25	96/256	0.6 MB	74.3 M	409.2	11.63
Conv3	3×3	5×5	256/512	1.1 MB	10.6 M	529.7	1.28
Conv4	3×3	3×3	512/512	2.3 MB	2.36 M	95.8	1.58
Conv. INF		**Avg. Power**		**167.5 mW**	**124.0 M**	**355.7**	**22.32**
		Avg. Efficiency		**2.12 – 4.52 TOPS/W (@ 0.78 – 1.1 V)**			
FC5	1×1	1×1	512/512	512 KB	795 K	151.7	3.43 [1)]
FC6	1×1	1×1	512/512	512 KB	534 K	106.4	3.28 [1)]
FC7	1×1	1×1	512/2	2 KB	2.08 K	106.2	0.01 [1)]
Online Learning		**Avg. Power**		**168.0 mW**	**1.33 M**	**129.5**	**6.71 [1)]**
		Avg. Efficiency		**0.771 – 1.36 TOPS/W (@ 0.7 – 1.1 V)**			

1) Five-iteration Learning

Fig. 5.21 Performance benchmark on MDNet [41]

RT-MDNet (640 × 480)	Kernel Size	Input Size	In/Out Channels	Params.	Total MAC	Effective GOPS	Avg. Time [ms]
Conv1	7×7	640×480	3/96	13.8 KB	1064 M	305.9	6.96
Conv2	5×5	160×120	96/256	0.6 MB	2821 M	605.4	9.32
Conv3	3×3	80×60	256/512	1.1 MB	5337 M	584.9	18.25
Conv4	3×3	3×3	512/512	2.3 MB	75.52 M	95.8	1.58
Conv. INF		**Avg. Power**		**163.2 mW**	**18.60 G**	**515.0**	**36.11**
		Avg. Efficiency		**3.16 – 6.74 TOPS/W (@ 0.78 – 1.1 V)**			
FC5	1×1	1×1	512/512	512 KB	381.6 M	145.9	10.27 [2)]
FC6	1×1	1×1	512/512	512 KB	256.3 M	111.0	9.83 [2)]
FC7	1×1	1×1	512/2	2 KB	31.2 K	110.8	0.04 [2)]
Online Learning		**Avg. Power**		**164.0 mW**	**638.9 M**	**129.5**	**21.01 [2)]**
		Avg. Efficiency		**0.790 – 1.39 TOPS/W (@ 0.7 – 1.1 V)**			

2) Fifteen-iteration Learning

Fig. 5.22 Performance benchmark on RT-MDNet [131]

[116] as the backbone network, and the corresponding performance was evaluated in the Udacity dataset [134] and the self-test dataset with VGA resolution (Fig. 5.22). The DF-LNPU achieved 38.5-ms DroNet inference latency at VGA resolution input. In addition, 48.5-ms learning latency was achieved in fine-tuning the last 5 layers for adaptations (Fig. 5.23).

Figure 5.24 shows the comparison table of the previous DNN learning processors [14, 15, 21, 33, 123, 128]. Compared with the previous FCL optimized learning processors [123, 128], our processor achieves 2.67 × higher energy-efficiency. Even though the processor [33] can support the tuning of quantization results, our processor supports BP and DFA method, so it strongly guarantees the DNN training accuracy. Our processor also shows better efficiency compared with commercial products [14, 15]. Specifically, the DF-LNPU can achieve the highest throughput

AlexNet CIFAR10	Kernel Size	Input Size	In/Out Channels	Params.	Total MAC	Effective GOPS	Processing Time [ms]
Conv1	3×3	32×32	3/64	1.7 KB	0.47 M	169.2	0.006
Conv2	3×3	8×8	64/192	0.1 MB	7.08 M	187.4	0.076
Conv3	3×3	4×4	192/384	0.6 MB	10.6 M	50.5	0.420
Conv4	3×3	4×4	384/256	0.8 MB	14.2 M	50.6	0.559
Conv5	3×3	4×4	256/256	0.6 MB	9.44 M	50.8	0.372
Conv. INF		**Avg. Power**		**127.1 mW**	**41.8 M**	**58.3**	**1.433**
		Avg. Efficiency		**0.46 – 0.98 TOPS/W (@ 0.78 – 1.1 V)**			
FC5	1×1	1×1	1024/512	1024 KB	524 K	51.2	0.66
FC6	1×1	1×1	512/512	512 KB	262 K	51.2	0.33
FC7	1×1	1×1	512/10	2 KB	1.02 K	42.7	0.01
FC INF		**Avg. Power**		**140.0 mW**	**787 K**	**51.1**	**0.99**
		Avg. Efficiency		**0.365 – 0.642 TOPS/W (@ 0.7 – 1.1 V)**			

Fig. 5.23 Performance benchmark on AlexNet [93]

	TETC 2017 [3]	GTX 1080 Ti [4]	TPU-v2 [5]	TCAS-I 2018 [6]	S. VLSI 2018 [7]	ISSCC 2019 [8]	This Work
Purpose	FC, RNN	GPU	CNN, FC, RNN	CNN, FC, RNN	CNN, FC	CNN, FC, RNN	**CNN, FC, RNN**
Training Support	BP	BP, DFA	BP	BP, FA	Quant. Tuning	BP	**BP, DFA, PDFA**
Technology [nm]	130	16	20	65	65	65	**65**
Die Area [nm^2]	13.96	471	-	3.52	12	16	**16**
Supply Voltage [V]	1.2	-	-	1.2	0.67 – 1.1	0.78 – 1.1	**0.7 – 1.1**
Frequency [MHz]	200	1480	-	200	20 – 200	50 – 200	**25 – 200**
Weight Precision	FXP8, 16	FP32	bfloat16	FXP16	FXP8	FP8, 16	**LC: FXP16 / ZCC: FP8**
Activation Precision	FXP8, 16	FP32	bfloat16	FXP13	FXP8	FP8, 16	**LC: FXP13 / ZCC: FP8**
Max Throughput [GOPS or GFLOPS]	38.4	10,157	45,000	51.2	5,638	FP16: 300[1)] – 1,002.1[2)] FP8: 600[1)] – 2,004.2[2)]	**LC: 155.2 ZCC: 300[1)] – 1,002.1[2)]**
FC Training Throughput/BW	-	20.49 OP/KB	73.24 OP/KB	32 OP/KB	-	-	**80.94 OP/KB**
Power Consumption [mW]	2,210	250,000	225,000	126	20.5 – 248.4	43.1 – 367	**LC: 17.9 – 252.4 ZCC: 9.3[2)] – 171.6[1)]**
Energy Efficiency [GOPS or GOPS/W]	17.38	40.63	200	406.4	411 – 62,100	FP16: 817[1)] – 15,600[2)] FP8: 1,630[1)] – 25,300[2)]	**LC: 614.9 – 1083.8 ZCC: 1,748[1)] – 27,072[2)]**

1) @ Sparsity: 0%, 2) @ Sparsity: 90%

Fig. 5.24 Performance comparison table

per bandwidth, 80.94 OP/KB. The main reason is that PDFA can reduce EMA and enable parallel processing by overcoming the BL problem.

Figure 5.25 depicts the demonstration system of the proposed DF-LNPU. We selected the object tracking application which is one of the on-chip learning scenarios requiring high speed and high energy-efficiency. The network and algorithm used for the demonstration system follow MDNet [41], which shows the best accuracy in the object tracking tasks. Although it outperforms the other object tracking networks, it has been considered hard to accelerate because of the heavy computation during the online training. The DF-LNPU successfully realizes the algorithm with the 44.8 FPS framerate.

The previous processor [123] also tried the real-time object tracking with the MDNet, but it used only two iterations for online learning due to the slow on-chip learning speed. It compensated for the accuracy degradation by network training for every video frame. However, it rather wasted the energy and induced overfitting which can be the other reason for accuracy degradation. In contrast, the DF-LNPU adopts five-iteration online learning, and the learning can be triggered only when the accuracy of the network is low.

Fig. 5.25 Object tracking demonstration system with DF-LNPU

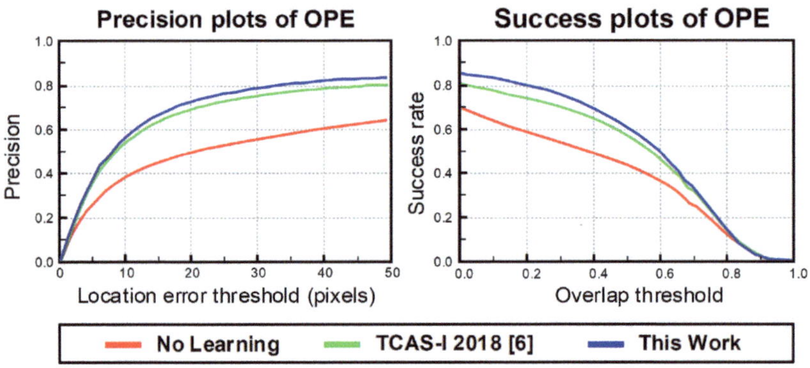

Fig. 5.26 Comparison of precision and success plots in OTB dataset [107]

We examined the performance of the previous processor [123] and the DF-LNPU by using the same object tracking benchmark (OTB) dataset [107]. Figure 5.26 shows both the precision and the success rate of one-pass evaluation (OPE). As depicted in the graph, both previous processor [123] and DF-LNPU show much higher accuracy compared with no online learning case. Furthermore, the DF-LNPU also outperforms the previous processor [123] thanks to the five-iteration triggered learning.

Figure 5.27 shows examples of the MDNet results. The DF-LNPU was demonstrated with both the public OTB dataset and the custom collected dataset. The DF-LNPU shows the best object tracking performance, but the previous processor [123] easily follows the wrong object which has a similar shape compared with the target objects (Fig. 5.28).

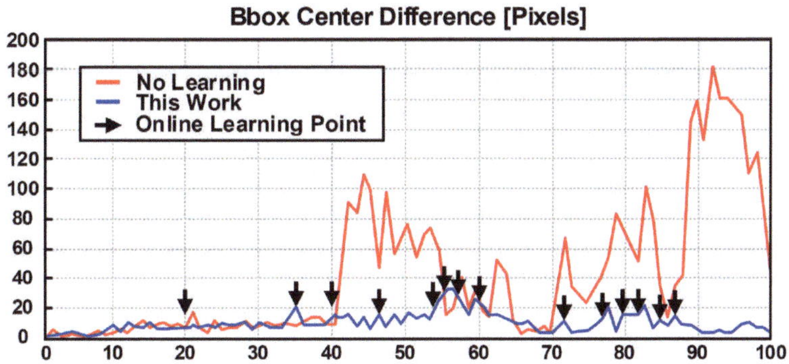

Fig. 5.27 The measurement of the bounding box (Bbox) center difference and online learning trigger point (Girl2 in OTB dataset)

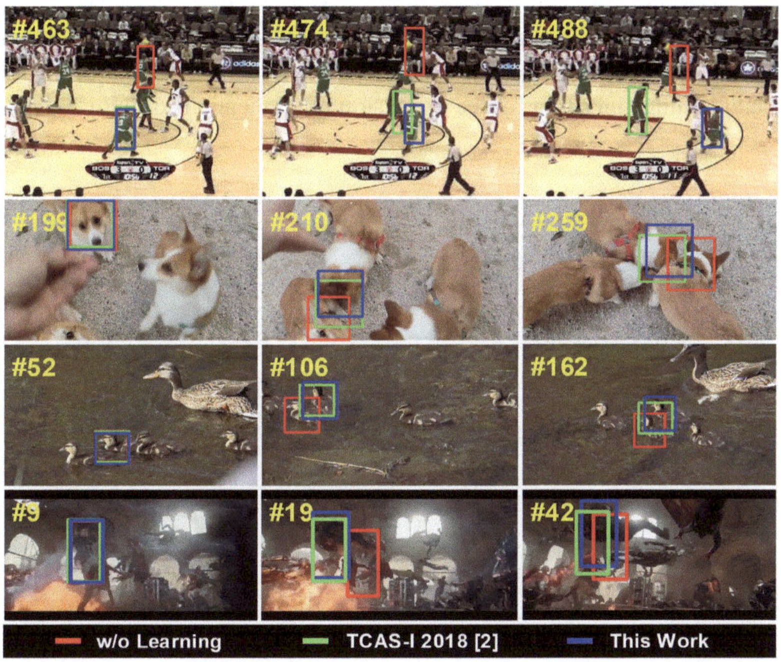

Fig. 5.28 Example Sequences of the Demonstration System. Tested in both public and custom datasets: basketball (OTB dataset), Corgi, Duck, and Ultron

5.7 Conclusion

The DF-LNPU, an energy-efficient DNN learning processor, realizes high-speed on-chip learning in mobile devices. The DF-LNPU implements both the zero-skip convolution core and PDFA based parallel learning processing core to reduce the overall processing time by 47.4%. The proposed processor solves BL problem by realizing PDFA based inter-core and intra-core pipeline, and it achieves 2.53 $\times$ higher throughput per bandwidth compared with the conventional low-power DNN learning processor [123]. The processor adopts the heterogeneous core architecture and shows at least 2.67 $\times$ higher energy-efficiency during the FCL learning scenario. Lastly, the DEPC with the PRNG encodes the BW with the minimum hardware cost and it can achieve the overall EMA reduction by 42.8%. In conclusion, our proposed DNN learning processor successfully demonstrates the fast online learning scenario with the highest energy-efficiency.

The DF-LNPU focused on real-time online DNN training, and it shows a typical example of an object tracking scenario. However, the current researches try to avoid online learning during object tracking and it sometimes follows the Siamese network for efficient object tracking. As shown in the measurement result, the DF-LNPU has heterogeneous architecture, and half of the design becomes idle to accelerate Siammask [132] based tracking. In further work, we will focus on a homogeneous core approach to achieve high efficiency in both online learning and Siamese network-based object tracking networks. In addition, we will develop a DFA based federated learning algorithm for the expansion of applications of DF-LNPU.

Chapter 6
HNPU-V1: An Adaptive DNN Training Processor Utilizing Stochastic Dynamic Fixed-Point and Active Bit-Precision Searching

This chapter presents HNPU, which is an energy-efficient DNN training processor by adopting algorithm–hardware co-design. The HNPU supports stochastic dynamic fixed-point representation and layer-wise adaptive precision searching unit for low-bit-precision training. It additionally utilizes slice-level reconfigurability and sparsity to maximize its efficiency in both DNN inference and training. Adaptive bandwidth reconfigurable accumulation network enables reconfigurable DNN allocation and maintains its high core utilization even in various bit-precision conditions. Fabricated in a 28-nm process, the HNPU accomplished at least 5.9 × higher energy-efficiency and 2.5 × higher area-efficiency in actual DNN training compared with the previous state-of-the-art on-chip learning processors.

This chapter is based on papers of "HNPU: An Adaptive DNN Training Processor Utilizing Stochastic Dynamic Fixed-Point and Active Bit-Precision Searching" (IEEE Journal of Solid-State Circuits [29]) and "A Mobile DNN Training Processor with Automatic Bit-precision Search and Fine-grained Sparsity Exploitation" (IEEE Micro [145]).

6.1 Introduction

Training deep neural network (DNN) requires a significant amount of computation which is possible only on cloud servers. For this reason, most of the DNN accelerators [1, 3, 4, 6, 7, 33] for edge or mobile applications have supported only DNN inference. Nowadays, on-chip learning processors for edge or mobile applications are receiving increasing attention due to their ability to various applications.

Firstly, it has a key role in distributed learning. Server-centric DNN training has a limited network scale-up, but it can be resolved by spreading the training workload across multiple edge devices. Moreover, federated learning [82], one of the typical

© The Author(s), under exclusive license to Springer Nature Switzerland AG 2023
D. Han, H.-J. Yoo, *On-Chip Training NPU - Algorithm, Architecture and SoC Design*,
https://doi.org/10.1007/978-3-031-34237-0_6

distributed learnings, can protect privacy by avoiding the upload of the user's private data to the server.

The second application is personalization which can provide user-specific services. The on-chip learning processors fine-tune a pre-trained network to recognize local data of the individual user environment. It can increase the recognition accuracy and also recognize new objects or categories which are customized by the user.

Lastly, on-chip learning can be an energy optimization method by continuously compensating for accuracy degradation. Mobile-oriented DNNs show less computation amount and memory access. However, they can be used in limited applications and show poor performance compared with the huge-size DNNs. On-chip learning can give mobile-oriented networks the opportunity to be utilized in practical situations. For example, a previous on-chip learning processor [123] utilized a light object recognition network, but it could have the object tracking functionality by adopting online DNN training. In this case, DNN training compensated for accuracy degradation that occurred by deformation of object shape or illumination changes during the tracking. The other example is temporal knowledge distillation (TKD) [42] which improves the detection performance by using online learning. Since it rarely uses the huge teacher network and utilizes a small student network most of the time, it gives another energy-efficiency improvement solution to the DNN system (Fig. 6.1).

Even though on-chip learning has big advantages, the implementation of an on-chip learning processor on the edge and mobile platforms is extremely challenging. Figures 6.2 and 6.3 explain two main difficulties that should be solved to realize the DNN training in mobile devices.

Firstly, the on-chip learning needs high throughput (>3TOPS) because it requires not only feed-forward (FF) but also error propagation (EP) and weight gradient (WG) stages. Generally, the DNN training is based on mini-batch gradient descent, and it utilizes multiple inputs as a batch and repeats the three training stages over multiple iterations. Therefore, the required amount of operations for the DNN training can be exponentially increasing according to the number of batches and iterations. Most on-chip learning processors [21, 26, 27, 138] concentrated to

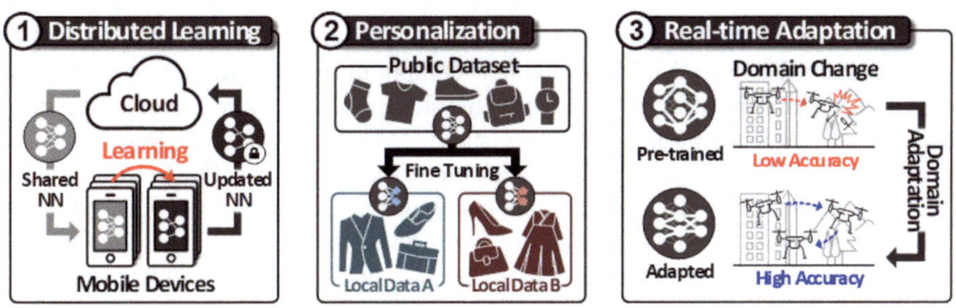

Fig. 6.1 Typical applications of DNN training on mobile devices

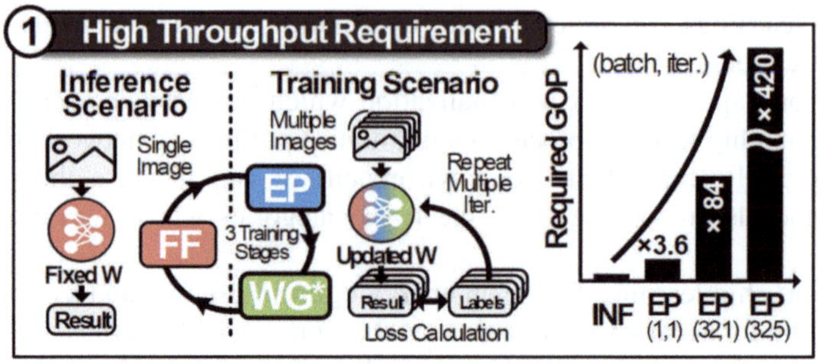

Fig. 6.2 Main difficulties of on-chip DNN learning: high throughput

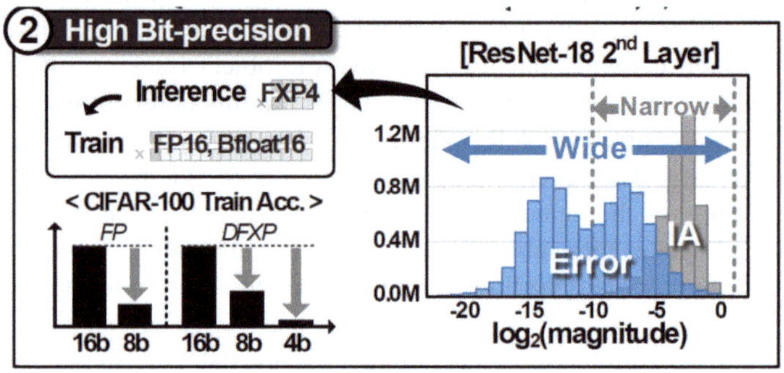

Fig. 6.3 Main difficulties of on-chip DNN learning: high bit-precision

improve the throughput by skipping zeros induced by the ReLU activation function, but the effect is minor because only 40–60% sparsity is appeared by ReLU in practical usages.

Secondly, DNN training needs high bit-precision. Unlike the input activation (IA) that appeared in the FF stage, the range of back-propagation error is much wider, so high-bit-precision error representation is necessary for accurate training. Since the DNN training is sensitive to the accuracy degradation, at least 16-bit floating-point (FP) representation has been considered for the previous on-chip learning processors. The high-bit-precision requirement of the DNN training results in low energy-efficiency for the previous on-chip learning processors. The LNPU [21] suggests FP16-FP8 mixed-precision computing to release this problem, but it still needs high-bit-precision multiply-and-accumulation (MAC) units. The area- and energy-efficiency of the mixed-precision FP MAC unit are much lower than the previous fixed-point (FXP) based bit-scalable processors [6, 7] (Fig. 6.4).

The proposed processor, HNPU, adopts algorithm–hardware co-optimization to achieve energy-efficient DNN training by solving these difficulties. The HNPU enables low-bit-precision DNN training by introducing a new number representation, stochastic dynamic FXP (SDFXP). It can find the optimal precision by introducing layer-wise adaptive precision scaling (LAPS). Its main core adopts bit-

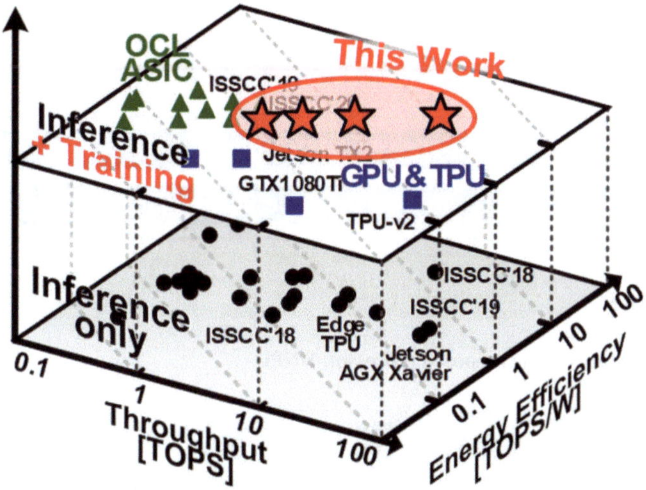

Fig. 6.4 Main difficulties of on-chip DNN learning: comparison with conventional DNN processors

slice serial architecture to minimize useless computations during the LAPS based DNN training. Even though a certain layer requires high bit-precision, both sparsity-aware input slice skipping (ISS) and precision-aware output slice skipping (OSS) techniques compensate for throughput and energy-efficiency degradation. Lastly, an adaptive bandwidth reconfigurable accumulation network (AB-RAN) gives DNN allocation reconfigurability while maintaining high core utilization even in various bit-precision conditions.

We organized the rest parts of the paper as follows. In Sect. 6.2, conventional low-bit-precision training methods are summarized and new low-bit-precision training algorithms, SDFXP and LAPS, are introduced in Sect. 6.3. Section 6.4 describes the overall architecture of the HNPU. Section 6.5 introduces three architecture and circuit-level features: (1) the bit-slice serial architecture with the SDFXP and LAPS units, (2) in–out slice skipping (IOSS) core architecture, and (3) AB-RAN. Section 6.6 shows chip implementation results, and the chapter will be concluded in Sect. 6.7 (Fig. 6.5).

6.2 Conventional Low-Bit-Precision Training Methods

Before explaining the details of HNPU, conventional low-bit-precision training methodologies will be summarized. We will deal with FP and FXP based DNN training and related low-bit-precision training methodologies.

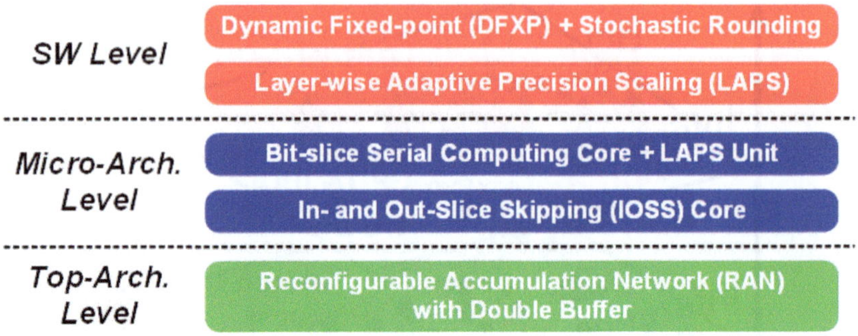

Fig. 6.5 Key features of the proposed processor: HNPU

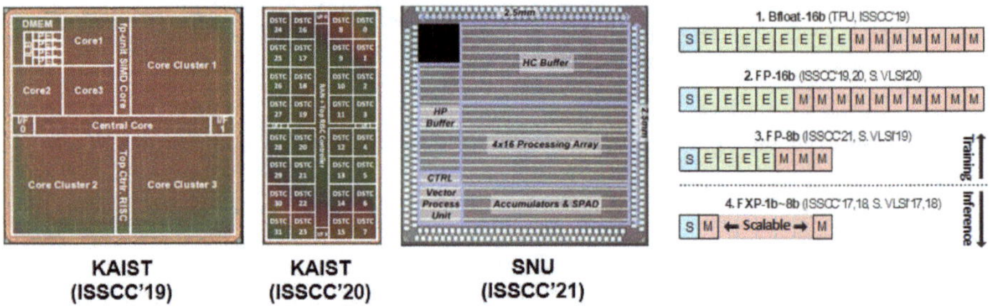

Fig. 6.6 Conventional DNN training processors using floating-point representation

6.2.1 Floating-Point Based Training

Conventional DNN training processors generally adopted FP based training because of its wide numeric range. Both LNPU [21] and GANPU [138] adopted FP16 training, but LNPU also utilized FP8 for energy-efficient DNN training. The LNPU adopted a fine-grained mixed-precision scheme to represent outliers with a sparse FP16 number format. There were also some trials to suggest new FP representations such as bfloat16 [14], Flexpoint [55], and DLfloat [18] to support energy-efficient on-chip learning. Additionally, a recent algorithm [56] insisted that FP8-only DNN training but accumulation was still done with high bit-precision such as FP16 (Fig. 6.6).

6.2.2 Fixed-Point Representation Based Training

FXP based computing has not been considered for training because of its limited numeric range and resolution. There were some papers [23, 27, 35, 120, 123] that realize on-chip learning with FXP representation, but it still requires almost 16-bit for scratch training [59], and low-bit-precision ($\leq$8-bit) training is only possible

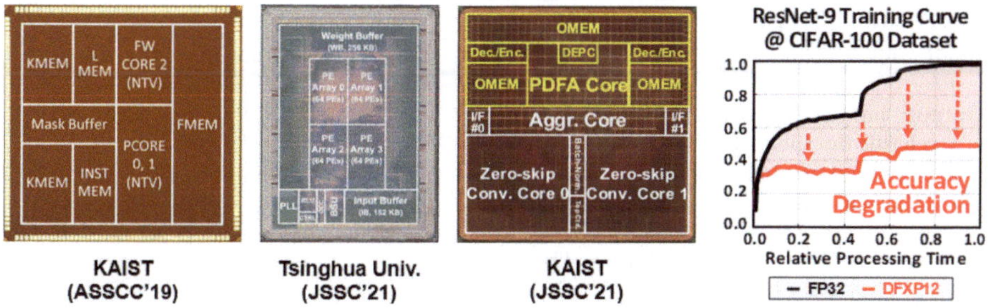

Fig. 6.7 Conventional DNN training processors using fixed-point representation

in the fine-tuning scenarios. Since the FXP based DNN training required higher bit-precision compared with the FP representation, most of the on-chip learning processors adopted FP based computing. It makes the on-chip learning processors unable to utilize the architecture of the conventional FXP based energy-efficient DNN inference processors such as look-up table [4] or bit-serial computing [7].

Currently, FXP based low-bit-precision training methods are developed based on stochastic rounding (SR [59]). The SR reduces rounding errors with probabilistic behavior. In other words, it can increase virtual representation resolution according to probabilistic rounding. Additionally, the primal weight, the original weight before bit truncation, should be maintained as high bit-precision. The bit truncation only occurs during the FF and EP stages, but WG is performed with high bit-precision to update primal weight. Both DoReFa-NET [135] and FxpNet [136] successfully demonstrated DNN training with their custom DNNs, but they lost generality because the network modification was necessary for both methods (Fig. 6.7).

6.3 Proposed Low-Bit-Precision Training Method

We introduced conventional low-bit-precision training methodologies, but they still required more than 8-bit to train general DNNs from the scratch. The proposed processor adopts FXP based new number representation for the low-bit-precision training. Moreover, the HNPU adopts an active precision searching algorithm, which is the first trial to find optimal precision during the DNN training.

6.3.1 Stochastic Thresholding and Stochastic Dynamic Fixed-Point Representation

The HNPU adopts the previous low-bit-precision training scheme, SR, and primal weight. Equation 6.1 represents the SR operation.

Algorithm 1 Stochastic thresholding

Input: input vector $\mathbf{x}$, current integer length $\mathbf{i_t}$, bit-width $\mathbf{B}$, pre-determined overflow threshold Th

Output: integer length of next iteration, $\mathbf{i_{t+1}}$

```
Stochastic threshold, Th_S = Th * Uniform(0, 1)
if Overflow_rate(x, i_t, B) ≥ Th_S then
    i_{t+1} = i_t + 1
else
    if Overflow_rate(x, i_t − 1, B) < Th_S then
        i_{t+1} = i_t - 1
    else
        i_{t+1} = i_t
    end if
end if
```

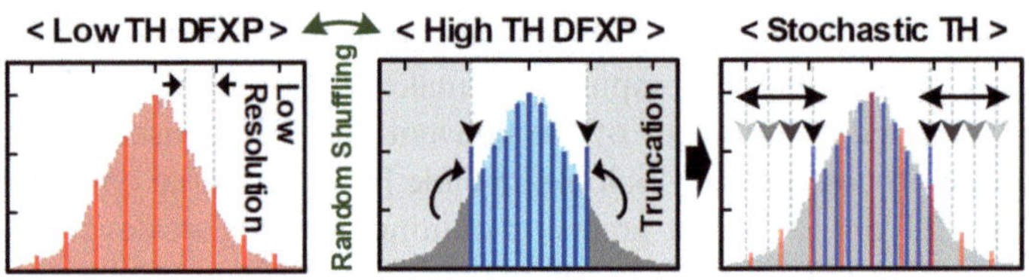

Fig. 6.8 Outline of stochastic thresholding

$$SR(x) = \begin{cases} \lfloor x \rfloor + 1 & \text{with probability } x - \lfloor x \rfloor \\ \lfloor x \rfloor & \text{with probability } 1 - (x - \lfloor x \rfloor) \end{cases} \tag{6.1}$$

Although the SR increases virtual representation resolution, the numeric range is still limited due to low bit-width. The DFXP [3] representation resolves this problem by controlling its fraction length according to layer-wise distributions. In the DFXP based computing, it counts overflow and decreases fraction length if the overflow appears. However, the distribution of the IA and weight is zero-centric, and only a minor portion of data has a large value and placed in the outlier of the distribution. Thus, many MSB bits are wasted to represent most small data. As shown in Fig. 6.8, strict fraction length control can precisely represent outlier data, but it can lose the DNN training accuracy because of inaccurate data representation that appeared in the majority of small data. To overcome this problem, fraction length can also be controlled only when the number of overflows is over the pre-determined threshold, but it can also be another reason for accuracy degradation because the outlier values can significantly affect the convolution results.

As shown in Algorithm 1, the HNPU solves this problem by introducing stochastic thresholding (ST). The ST randomly determines the threshold value to control fraction length. If the overflow occurs, the convolution results are clipped to the nearest value the current bit-precision can represent. The ST can enlarge the numeric

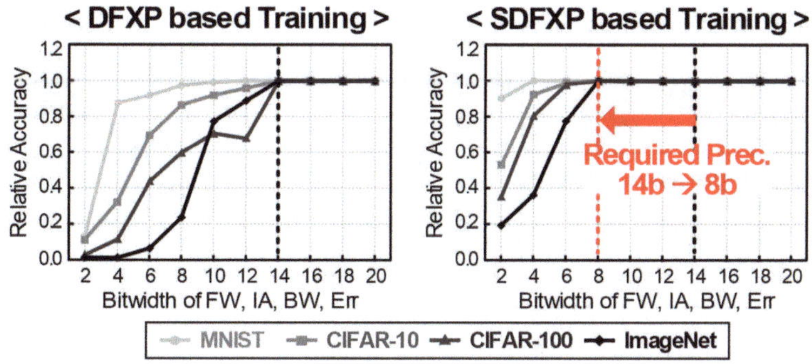

Fig. 6.9 SDFXP based bit-precision reduction result

representation range while minimizing the errors caused by clipping compared with the DFXP. Specifically, ST can compensate for the accuracy degradation when the training is done with less than 6-bit. Finally, a new number representation, stochastic DFXP (SDFXP), is proposed by combining three different concepts, DFXP, SR, and ST (i: integer length and f: fraction length).

$$SDFXP(x, i, f) = \begin{cases} 2^i - 2^{-f} & x \geq 2^i - 2^{-f} \\ 2^{-f} - 2^i & x \leq 2^{-f} - 2^i \\ 2^{-f} * SR(2^f x) & \text{otherwise} \end{cases} \tag{6.2}$$

As depicted in Fig. 6.9, we compared the ResNet-9 training performance of DFXP and SDFXP in four different datasets. It shows that the required bit-precision is reduced from 10-bit to 4-bit in the MNIST and 14-bit to 8-bit in the ImageNet dataset compared with the DFXP based training.

6.3.2 Layer-wise Adaptive Precision Scaling (LAPS)

According to Fig. 6.9, the SDFXP successfully reduces the required bit-precision for DNN training. Furthermore, it shows that the required bit-precision can be varied according to the target dataset even if it utilizes the same network architecture. In spite of this characteristic, conventional on-chip learning processors [14, 18, 21, 23, 26, 27, 35, 120, 123, 138] used fixed high bit-precision to train the DNNs because optimal precision searching has been considered impossible during the DNN training phase. Therefore, active bit-precision searching during the training can be the key functionality for the energy-efficient design of the on-chip learning processor.

The proposed processor supports automatic bit-precision searching by adopting a new training method, LAPS. Figure 6.10 and Algorithm 1 show the outline of

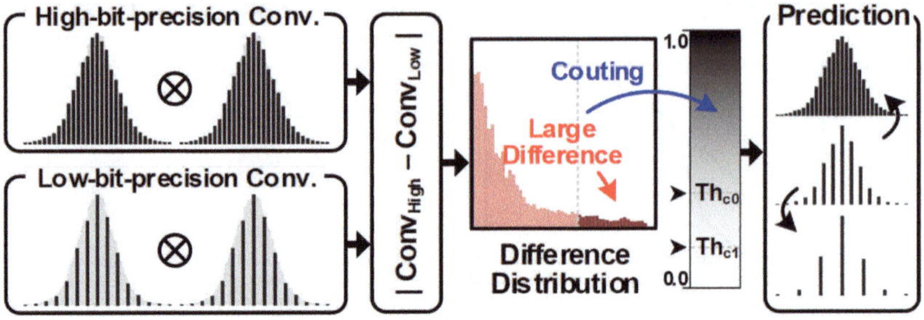

Fig. 6.10 Outline of layer-wise adaptive precision scaling

Dataset	Network	FP16		FP8		SDFXP12		SDFXP8		SDFXP4		LAPS			
		Train Acc.	Test Acc.	Train Acc.	Test Acc.	Train Acc.	Test Acc.	Train Acc.	Test Acc.	Train Acc.	Test Acc.	Train Acc.	Test Acc.	Avg. Prec.	Weighted Avg. Prec.
MNIST	ResNet-9	99.90	99.22	99.28	98.78	99.82	99.15	99.84	99.21	99.32	99.01	99.40	99.02	4.32	4.41
	ResNet-18	99.73	99.18	99.60	99.22	99.71	99.17	99.75	99.21	99.30	99.05	99.14	98.88	4.10	4.24
	ResNet-50	99.56	98.82	99.44	99.08	99.66	99.12	99.57	99.25	99.08	99.04	99.24	99.03	4.20	4.38
	VGG-16	99.65	99.06	99.58	98.99	99.67	99.23	99.64	99.07	99.59	98.32	99.04	98.70	4.00	4.00
	SENet	99.75	99.20	99.63	99.10	99.68	99.06	99.70	99.13	99.20	98.78	99.22	98.83	4.51	4.60
CIFAR 10	ResNet-9	99.86	93.69	84.27	83.54	99.90	93.87	99.86	93.73	90.12	86.58	98.58	92.94	5.84	6.27
	ResNet-18	99.98	95.22	99.94	93.23	99.98	95.20	99.98	95.01	90.44	85.68	97.88	92.48	5.14	5.53
	ResNet-50	99.97	94.64	99.91	93.84	99.96	95.18	99.98	95.17	88.53	81.34	99.88	94.28	6.02	8.00
	VGG-16	99.95	93.82	99.90	92.86	99.94	93.41	99.93	93.69	89.22	86.02	99.78	93.11	6.25	6.61
	SENet	99.98	95.02	99.93	93.71	99.97	94.86	99.96	94.74	79.33	75.97	96.76	92.40	7.07	8.00
CIFAR 100	ResNet-9	98.17	73.14	26.83	21.08	98.82	73.48	98.64	73.27	65.27	57.51	97.57	72.31	6.53	7.07
	ResNet-18	99.96	77.33	99.92	73.44	99.97	78.37	99.96	78.18	83.74	62.85	98.14	75.02	5.70	6.15
	ResNet-50	99.92	78.04	64.47	48.82	99.96	79.01	99.95	79.28	78.74	60.44	99.35	74.22	6.10	8.00
	VGG-16	99.72	73.90	99.67	72.25	99.84	73.61	99.75	73.35	68.78	59.15	99.20	71.81	6.54	6.89
	SENet	99.94	77.33	99.90	75.72	99.96	76.71	99.95	76.63	79.58	62.31	99.65	75.16	7.24	8.00

Fig. 6.11 CNN training result with SDFXP and LAPS

the LAPS algorithm. The LAPS calculates the difference between low-bit-precision and high-bit-precision convolution results to determine whether the DNN requires higher or lower bit-precision. The difference calculation occurs only in the first three iterations per epoch. It increases the counting numbers if the difference is larger than the pre-determined threshold value. And then, the internal finite-state machine increases or decreases the bit-precision of the corresponding layer based on the counting results. Determined precision is applied from the next iteration and it is maintained until an epoch is finished. In summary, the LAPS indirectly decides layer-wise optimal precision by comparing the low-bit-precision and high-bit-precision convolution results, and the precision is changed once for each epoch.

Figure 6.11 shows the convolutional neural network (CNN) training results by using LAPS. We measured both the train and test accuracy of four different DNNs in three typical datasets. It uses 16-bit primal weight in both FP and SDFXP based DNN trainings. The SDFXP8 based training shows similar or slightly better accuracy compared with the FP16 training. However, the SDFXP4 based training sometimes fails to achieve original high accuracy. The LAPS automatically finds the optimal precision of each layer to minimize the required bit-precision during the DNN training. One key observation is that the required bit-precision should be larger

Dataset	Network	FP32		SDFXP12		SDFXP8		SDFXP4		LAPS			
		Train Acc.	Test Acc.	Train Acc.	Test Acc.	Train Acc.	Test Acc.	Train Acc.	Test Acc.	Train Acc.	Test Acc.	Avg. Prec.	Weighted Avg. Prec.
ImageNet 32×32	ResNet-9	59.02	47.79	58.90	47.23	57.07	46.62	31.94	26.35	56.90	46.47	7.90	7.95
	ResNet-18	84.41	53.52	82.96	53.44	62.45	50.40	41.21	30.82	79.31	53.51	11.13	11.74

Fig. 6.12 ImageNet (32×32 resolution) training result with SDFXP and LAPS

if the task becomes more difficult. In the MNIST training, the required precision is almost 4-bit, but it requires more than 6-bit during the CIFAR-100 training.

Figure 6.12 shows additional ImageNet (32×32 Resolution) training results with the SDFXP and LAPS. Although Fig. 6.11 simulation is done with 16-bit primal weight, it adopts 20-bit primal weight in ImageNet training. Consequently, the LAPS based training automatically finds the layer-wise optimal precision even in various datasets and network types.

6.4 Overall Architecture

We introduced a new number representation, SDFXP, and an active precision searching methodology, LAPS. The proposed processor, HNPU, adopts both SDFXP and LAPS for high-speed and energy-efficient DNN training (Fig. 6.13).

Figures 6.14 and 6.15 show the overall architecture of the HNPU which consists of thirty-two bit-slice training cores (BSTCs), a peripheral training assistant, two optimizer cores (OPTCs), and a TOP RISC controller. Each BSTC has four processing element (PE) rows, IA pre-fetchers, 4 KB IA memory, and buffers to store the weights and OAs. The entire memory included in the BSTC adopts double buffering to maintain high core utilization. The IA pre-fetcher includes 4-entry FIFOs which temporally store input data without zero-slices (ZSs) and only non-zero-slices (NZSs) are fetched to the PE rows. Accumulation is performed inside the PE rows. The aggregation module collects accumulation results to complete convolution or matrix multiplication operations. Final results are transferred to the OA buffer or accumulation switch (ACC SW). The ACC SW includes an SDFXP unit that performs post-processing to convolution results. It has the role of batch normalization (BN) and activation function computing which is an essential part of the DNN training. Every BSTC is connected with both the data NoC and the 2D mesh-type AB-RAN.

The peripheral training assistant consists of a LAPS unit, a workload monitor, and a channel reordering module with a 120 KB peripheral memory (PMEM). It is used as the temporal buffer to store the IAs or weights just before and after external memory access. It simultaneously monitors sparsity patterns of IA and reflects it for workload balancing. Moreover, the channel reordering unit performs a weight transpose operation during EP computing. The LAPS unit compares the low-bit-precision and high-bit-precision convolution results which are temporarily stored in the PMEM. The last unit, OPTC, can support three types of optimizers: SGD, momentum, and signADAM++ [139].

Algorithm 2 Layer-wise adaptive precision scaling

Input: total iteration number of an epoch $\mathbf{T}$, current iteration $\mathbf{tin[0, T-1]}$, input vector $\mathbf{x}$, weight $\mathbf{w}$, corresponding bit-width $\mathbf{B_{x,t}}$, $\mathbf{B_{W,t}}$, difference threshold $\mathbf{Th_d}$, count threshold $\mathbf{Th_{c0}}$, $\mathbf{Th_{c1}}$

Output: bit-width of the next iteration $\mathbf{B_{x,t+1}}$, $\mathbf{B_{W,t+1}}$

```
C_Low = Conv(x, w, B_x,t, B_W,t)
if t < 3 then
    if t = 0 then
        C_High = Conv(x, w, B_x,t + 2, B_W,t)
        if Count(|C_High − C_Low| > Th_d) > Th_c0 then
            B_x,t+1 = B_x,t + 1
        else
            if Count(|C_High − C_Low| > Th_d) < Th_c1 then
                B_x,t+1 = B_x,t − 1
            end if
        end if
    else
        if t = 1 then
            C_High = Conv(x, w, B_x,t, B_W,t + 2)
            if Count(|C_High − C_Low| > Th_d) > Th_c0 then
                B_W,t+1 = B_W,t + 1
            else
                if Count(|C_High − C_Low| > Th_d) < Th_c1 then
                    B_W,t+1 = B_W,t − 1
                end if
            end if
        else
            if t = 2 then
                C_High = Conv(x, w, B_x,t + 2, B_W,t + 2)
                if Count(|C_High − C_Low| > Th_d) > Th_c0 then
                    B_x,t+1 = B_x,t + 1
                    B_W,t+1 = B_W,t + 1
                else
                    if Count(|C_High − C_Low| > Th_d) < Th_c1 then
                        B_x,t+1 = B_x,t − 1
                        B_W,t+1 = B_W,t − 1
                    end if
                end if
            end if
        end if
    end if
end if
```

6.4.1 Main Training Cores

It consists of four clusters and each cluster includes eight main training cores. Each core computes partial tiles of the input activation (IA) and supports both convolution and matrix multiplication operations. It adopts bit-slice serial architecture (BSSA) which controls its computation cycles to support various bit-precisions. Each PE computes 4b $\times$ 4b at a cycle, but the MAC is designed with 5b $\times$ 5b because of the sign extension. It can compute high-precision computing by performing MAC

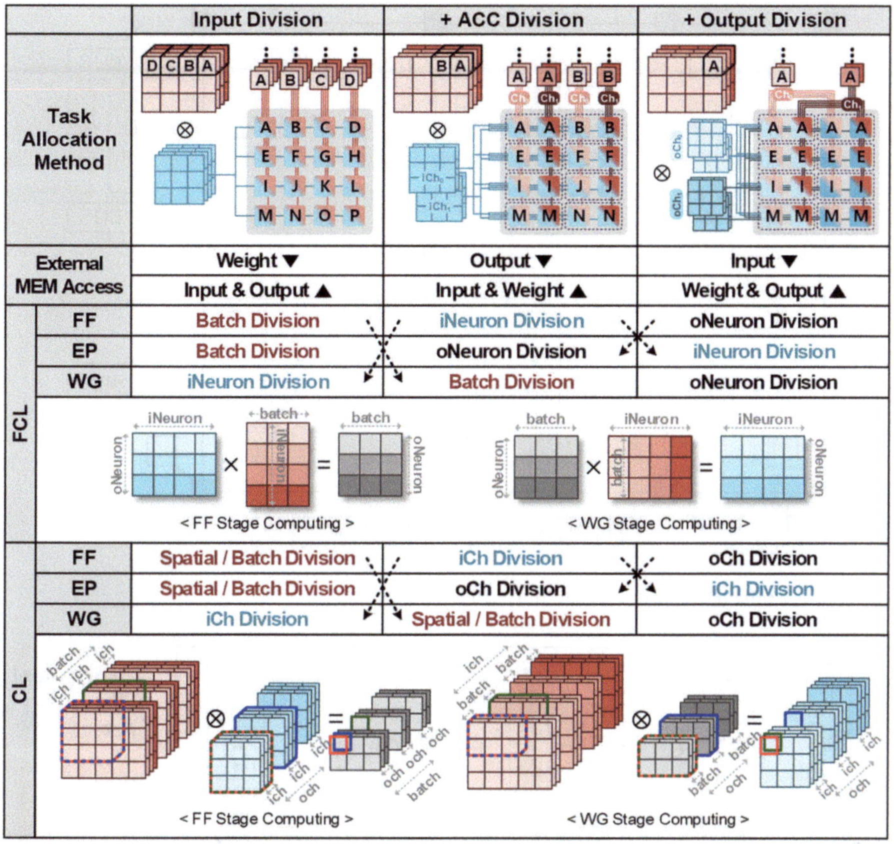

Fig. 6.13 Three task allocation methodologies during DNN training

operations over multiple cycles. Internal input slice skipping (ISS) pre-fetcher helps the processor to skip the zero-slice (ZS), and it adaptively controls the address of the IAMEM and weight buffer. Although the IA is broadcasted along with PE rows, the weights are unicasted to each PE array (PA). Both IAMEM and weight buffer adopt double-buffer to receive the next operands while the core is in use. Every core is connected with data NoC for free data transaction among the internal cores.

6.4.2 *Post-processing for Low-Precision Training*

A cluster also includes a local accumulation (ACC) core and it receives computation results from 8 cores. It controls the ACC path to support variable precisions, kernel sizes, and layer types. The proposed processor adopts a reconfigurable ACC network (RAN [138]) to give programmability. Each core is connected with the ACC switch

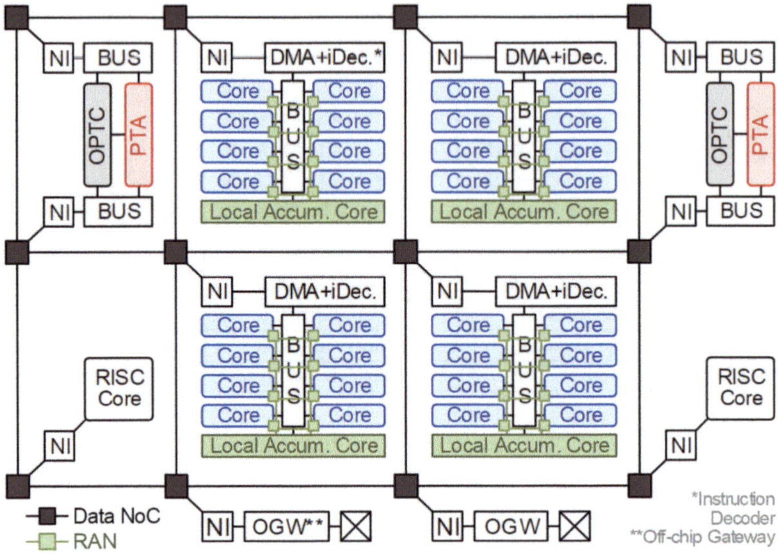

Fig. 6.14 Details of network-on-chip in HNPU

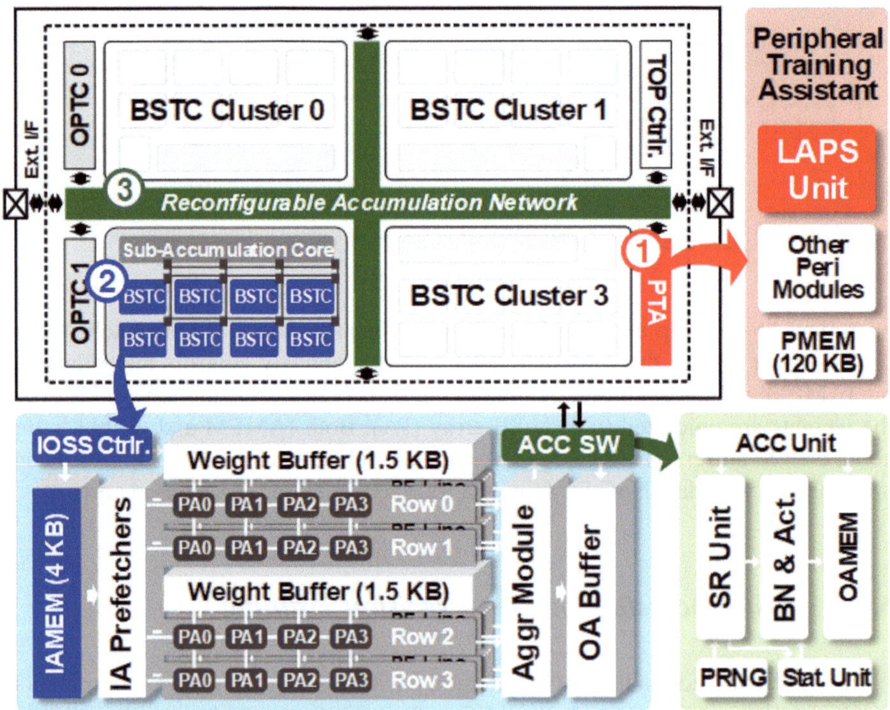

Fig. 6.15 Overall architecture of HNPU

(SW) and it performs post-processing of convolution output. The statistic module evaluates the mean and variance of the data to support batch normalization (BN). It also judges whether the overflow or underflow occurs. The proposed processor adopts DFXP representation and controls fraction length by utilizing these overflow and underflow information. The changed fraction length is reflected from the next iteration. Moreover, the ACC SW is co-designed with a unit that adds random noise generated by an internal pseudo-random number generator to the accumulation results. By implementing SR, the low-precision training is now possible. As a result of the experiment on the CIFAR-100 dataset, the conventional FXP based training shows only 2.5% training accuracy, but the proposed processor achieves 98.7% accuracy by utilizing both DFXP and SR.

6.4.3 Task Allocation Methodologies for Training

Conventional training processors [21, 34] insisted that a new hardware architecture supporting different data paths is needed to implement the three training stages: FF, EP, and weight gradient (WG). However, in this processor, the data path required for DNN training is unified by gathering batch data. Figure 6.13 summarizes three different task allocation methods of the proposed processor. For instance, on CNN training, there are three main elements to determine task allocation method: batch, input channel, and output channel. From the view of the FF stage, a different batch shares the same weight but generates different output. To complete one output, the processor needs to accumulate the multiplication results of the multiple input channels. Lastly, different output channel shares a single input but generates different outputs by using the different weights. In the EP and WG stage, the role of the batch, input channel, and output channel is interchanged as shown in Fig. 6.13. Thanks to the batch gathering, it can finally support the entire three training stages even with a single unified data path. Instead, the NoC is designed to support data broadcasting and it combines RAN to support accumulation among the multiple cores. This allocation method can also be utilized in FC layer training, but the only difference is that three elements are changed as a batch, input, and output neuron. The proposed processor uses a mix of the three methods in response to different memory access patterns for each layer.

6.4.4 Peripheral Components and Optimizer Core

In addition to the main training cores, the proposed processor contains a single peripheral training assistant (PTA) unit and two optimizer cores (OPTCs). The PTA contains a LAPS unit to support active precision search during the DNN training. Since the weight update process is a memory-intensive operation, an independent core called OPTC is designed to manage weight, gradient, and momentum data,

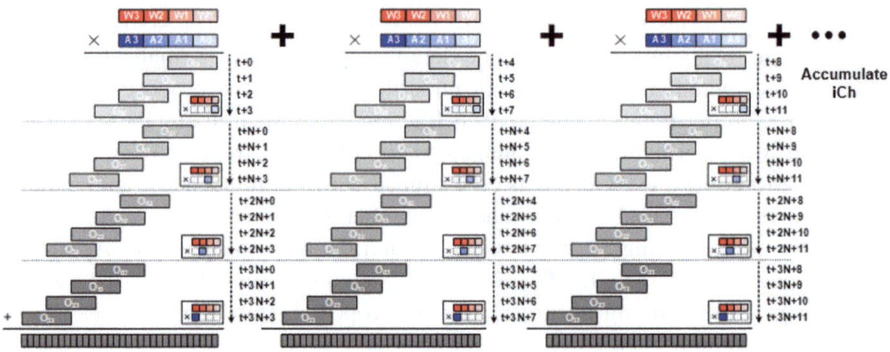

Fig. 6.16 Outline of bit-slice serial computing (red: weight and blue: input activation)

supporting three types of optimizers: SGD, Momentum, and ADAM. Specifically, it includes signADAM++ [139] unit to approximate ADAM optimizer with the minimum memory transaction. Thanks to the optimizer approximation, it reduces 40% external DRAM access compared with the original ADAM optimizer while maintaining its fast training curve. The PCRU is also placed in the OPTC and modifies weight order according to its sorting result. It has the role of workload balancing during the sparsity-aware acceleration, and the details of its operation will be covered in the later section (Fig. 6.16).

6.5 Key Features of the Proposed Processor

Section 6.3 introduced algorithm-level key features adopted in the HNPU. Architecture and circuit-level features will be described in this section. Firstly, the bit-slice serial architecture with the pseudo-random-number based SDFXP unit and LAPS unit will be explained. An explanation of energy-efficient IOSS core architecture together with AB-RAN will follow.

6.5.1 Bit-Slice Serial Architecture with SDFXP and LAPS Unit

Bit-Slice Training Unit with Bit-Slice Serial Computing

Figure 6.17 illustrates the detailed computation flows of the HNPU. The main DNN core, BSTC, includes a bit-slice training unit and an SDFXP unit. The bit-slice training unit uses 4-bit × 4-bit MAC as a base unit and receives a 4-bit-slice at a time after dividing high-bit-precision data into a number of 4-bit-slices. Conventional bit-serial architecture [7] used in many bit-reconfigurable processors requires high accumulation bit-width to prevent swamping problem [56]. The main reason is

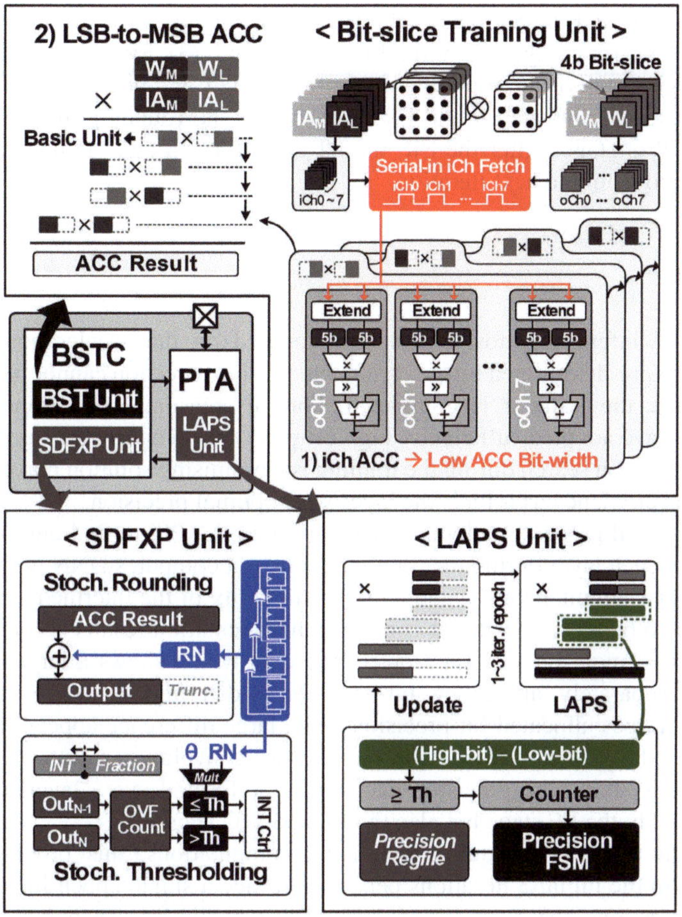

Fig. 6.17 Bit-slice training core with SDFXP and LAPS unit

that it computes LSB-to-MSB accumulation before it calculates further operands. To solve this problem, the HNPU computes input channel accumulation before the move to the MSB data. Since it firstly accumulates the LSB data, it prevents swamping problem even with the small bit-width. As a result, the proposed bit-slice serial architecture reduces accumulation bit-width by 38.1% thanks to the change of accumulation order.

SDFXP Unit with Pseudo-Random Number Generator

After the bit-slice training unit completes the convolution, final accumulation results are transferred to the SDFXP unit. The SDFXP unit includes a linear-feedback-shift-register-based pseudo-random number generator and the random number (RN) is used for the SR and the ST. The SDFXP unit can support three different rounding modes: (1) no-rounding, (2) normal-rounding, and (3) SR. Based on training methodologies, one of the rounding modes is selected. In our CNN training setup, the SR is used during the FF and EP stage, but the no-rounding mode is adopted during the WG stage. After the SR, the ST is performed and the updated integer length is used from the next iteration.

LAPS Unit

In the DNN inference, most NPUs only focused on the acceleration of DNN according to pre-determined bit-precision, but this bit-precision cannot be predicted easily before the training is completed. For this reason, conventional training processors have adopted sufficiently high bit-precision to ensure the accuracy. The run-time precision search during the training is a promising solution that can further improve the efficiency as it finds the layer-wise optimal precision.

The proposed processor adopts a new algorithm, LAPS, to find out the optimal precision of each layer during the training. At first, it computes DNN with high and low precisions simultaneously, and then the similarity of the intermediate output is computed. To be more specific, it compares the calculation results before and after allocating two more bits for each layer. The data which shows a large difference is counted separately, and a new bit-precision is determined based on the count result. It enlarges allocated bit-precision if the large difference appears frequently and reduces the precision if it judges that the two convolution results look similar. This similarity-based bit-precision decision is performed every layer and it also affects not only the FF stage but also the EP and WG stage.

Although the LAPS requires additional computations, this process is only performed in the initial 3 iterations per epoch. Since a single epoch includes more than 300 iterations, the overhead of the LAPS is negligible. Thanks to the LAPS algorithm, the proposed processor can find layer-wise optimal precision even though the training is in progress. From our training benchmark results, LAPS achieves FP16-level training accuracy in the CIFAR dataset while utilizing less than 8-bit precision. It can further improve its efficiency by enabling ultra-low-precision training if the target dataset is simple enough such as MNIST.

Figure 6.18 shows the LAPS unit placed in a PTA and it receives both original and higher precision computation results immediately from the main core. Since our main cores adopt BSSA, the overhead of difference calculation can be minimized by excluding MSB accumulation during the high-precision computing. Finally, the LAPS unit counts the large difference data and updates new bit-precision using the internal finite-state machine (FSM). The FSM decides the precision according to pre-determined counting threshold and bit-precision changing order. The BSSA controls operation cycles according to the varied precision found by LAPS unit and maximizes both throughput and efficiency. Figures 6.19 and 6.20 summarize the training results of the proposed processor after LAPS based optimization. The

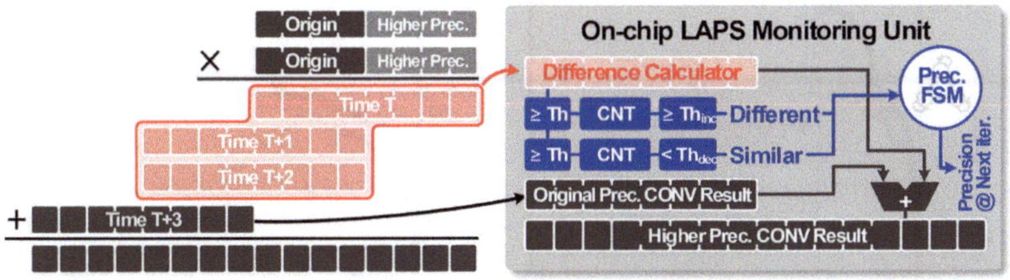

Fig. 6.18 On-chip LAPS monitoring unit with built-in finite-state machine: a combination of LAPS unit and BSSA for faster difference calculation

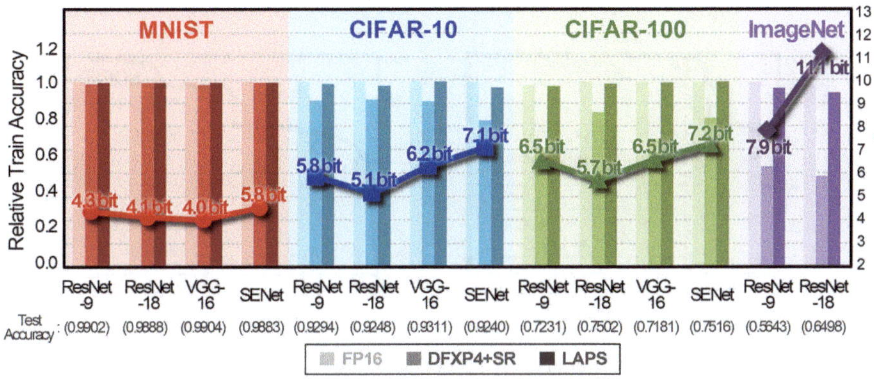

Fig. 6.19 CNN training results of HNPU with the help of LAPS: relative training accuracy and final test accuracy of LAPS based training with the average bit-precision

Fig. 6.20 Relative reduction of energy consumption after LAPS based DNN training

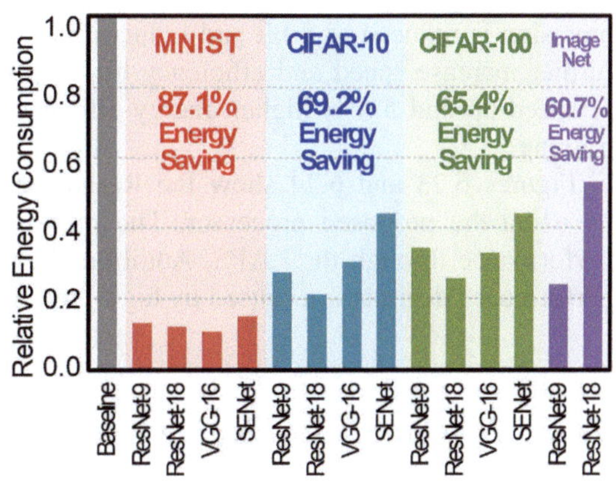

combination of the LAPS unit and BSSA finally reduces 77.3% processing time and 73.9% energy consumption compared with the DFXP based training.

Figures 6.21 and 6.22 show the training results with the BSTC including SDFXP and LAPS unit. Previously, the training of ResNet-9 and ResNet-18 requires more than 12-bit when it adopts DFXP. It spends 9 × longer time compared with the

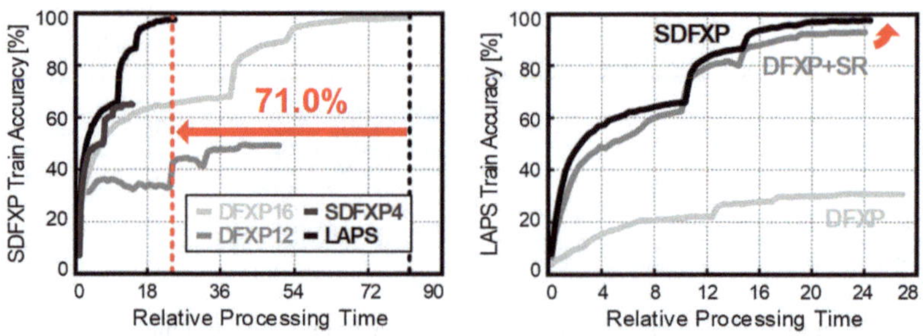

Fig. 6.21 CNN training result with proposed processor: ResNet-9 training result in CIFAR-100 dataset

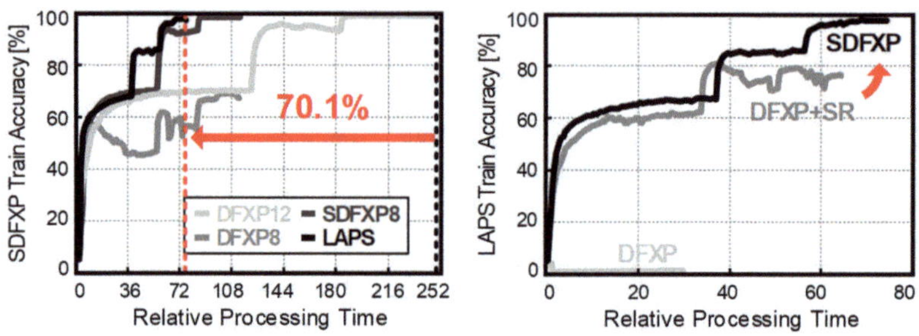

Fig. 6.22 CNN training result with proposed processor: ResNet-18 training result in CIFAR-100 dataset

4-bit training. When the HNPU adopts SDFXP representation, the required bit-precision is reduced to 8-bit and induces 55.6% processing time reduction. It can further increase speed and efficiency by adopting LAPS and it finally shows 3.4 × speed-up and 5.6 × higher energy-efficiency compared with the DFXP based training.

Figures 6.23 and 6.24 show the ResNet training example at the CIFAR-100 by using the proposed processor. The proposed processor optimizes layer-wise performance through the LAPS. Additionally, the ISS and PCRU compensate for performance degradation caused by high-precision computing.

6.5.2 In–Out Slice Skipping Core Architecture

Previous on-chip learning processors [21, 26, 27, 138] tried to utilize word-level sparsity to improve both throughput and energy-efficiency. Therefore, their performance highly depended on the sparsity caused by ReLU. Input sparsity exploitation such as LNPU is not efficient for DNN training because the EP does not induce input sparsity. Dual-sparsity-aware design can overcome this problem, but it cannot utilize both input and output sparsity simultaneously because only one of the

Layer Number	Kernel Size	IA Size	In/Out Ch.	Train Stage	Slice-level Sparsity	Avg. TOPS	Avg. TOPS/W	Avg. Bit-width FW/BW	Avg. Bit-width IA/Error
1	3×3	32×32	3/64	FF	9.0%	1.65	2.08	12.00	12.00
				EP	-	-	-	-	-
2	5×5	32×32	64/128	FF	74.3%	6.97	10.33	7.63	7.70
				EP	64.4%	5.94	7.71	7.54	7.63
3	3×3	16×16	128/128	FF	84.9%	12.28	21.98	4.11	4.49
				EP	95.8%	12.28	28.31	4.72	5.32
4	3×3	16×16	128/128	FF	76.8%	12.28	18.93	4.55	4.98
				EP	95.7%	12.28	28.23	4.37	4.85
5	3×3	16×16	128/256	FF	67.5%	6.25	8.42	7.37	7.63
				EP	73.3%	6.86	10.01	6.90	7.50
6	3×3	8×8	256/256	FF	86.3%	8.02	14.69	7.24	7.50
				EP	95.6%	7.97	18.73	7.16	7.49
7	3×3	8×8	256/256	FF	83.6%	7.85	13.61	6.67	6.81
				EP	96.6%	12.28	28.86	6.34	6.95
8	3×3	8×8	256/256	FF	83.3%	7.82	13.48	7.75	7.77
				EP	83.1%	7.81	13.42	6.57	7.16

Fig. 6.23 Detailed performance analysis on ResNet training

sparsity types, either input or output sparsity, appears during the training. Output zero prediction can be a solution to fully utilize both input and output sparsity at the same time, but it causes area and energy overhead when the prediction block becomes huge to achieve fast prediction.

Slice-Level Input Skipping

As shown in Fig. 6.25, the HNPU releases ReLU dependency by utilizing slice-level sparsity. The DNN usually shows Gaussian-like distribution in IAs after the BN layer. Among them, half of the IA is evaluated as zero because of the ReLU. In addition, the majority of data is concentrated near zero and they are significantly smaller than the maximum values of the distribution. In that case, near-zero data wastes MSB bits when using the FXP representation. It gives great potential to the slice-level sparsity exploitation. Since the slice-level input sparsity does not depend on ReLU, it can also appear during the EP stage. The HNPU can utilize more than 60% input slice sparsity in both FF and EP stages. Unlike the conventional dual-sparsity-aware hardware [26, 27, 138], the HNPU can achieve high throughput during the EP stage even without the output sparsity exploitation unit (Fig. 6.26).

The proposed processor minimizes its operation cycles according to the bit-precision determined by the LAPS algorithm. However, its performance can be significantly degraded when the required bit-precision becomes higher. Sparsity exploitation is one method to compensate for throughput degradation during high-precision computing. The first type of sparsity exploitation [21] is input zero

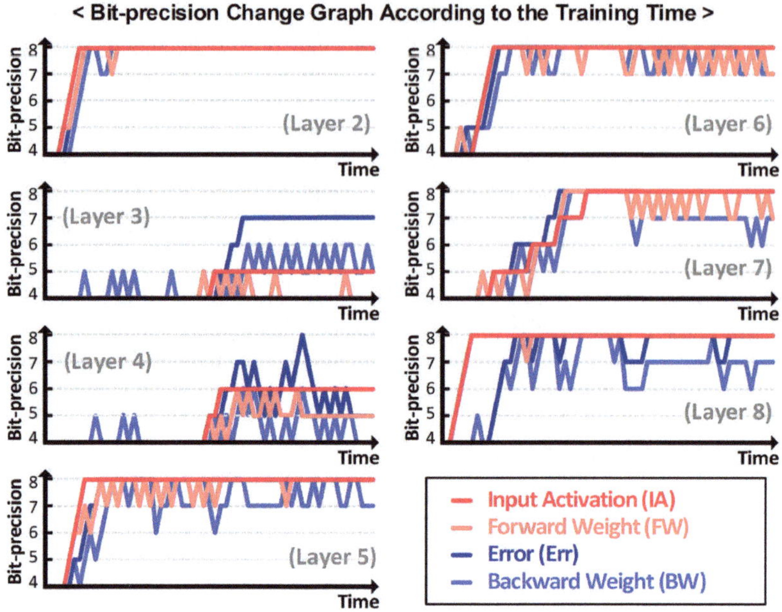

Fig. 6.24 Bit-precision change graph according to the DNN training time

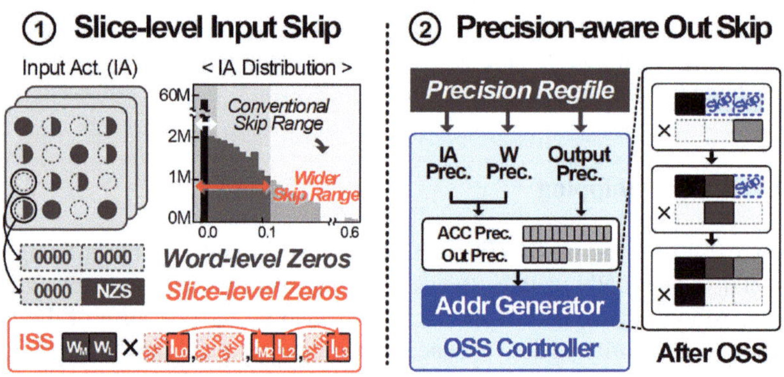

Fig. 6.25 Outline of in- and out-slice skipping

skipping which skips word-level zeros induced by the ReLU activation function. However, the input sparsity appears only in the FF stage and it is useless in the EP stage. Dual-sparsity exploitation [27, 138] can utilize zero skipping even in the EP stage but increases hardware complexity by supporting both input and output zero skipping simultaneously. In this case, the efficiency is rather reduced when the DNN uses the non-ReLU activation because the word-level sparsity completely disappears (Fig. 6.27).

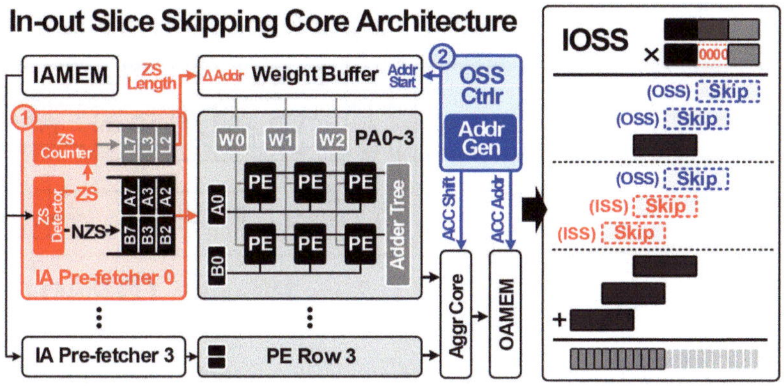

Fig. 6.26 IOSS core architecture

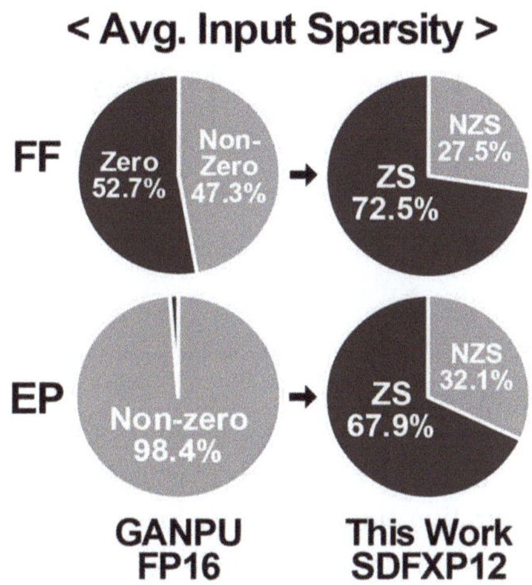

Fig. 6.27 Average sparsity of feed-forward and error propagation stage

To resolve this problem, the proposed processor exploits slice-level sparsity of the input operands. If the bit-width of a single slice is defined as 4-bit, we verified that more than 60% of ZS always appears during the ResNet training. The main reason is that the distribution of IA or error is concentrated at zero, so it can skip MSB zeros even without word-level sparsity. As shown in Fig. 6.28, our main core includes an ISS pre-fetcher which gives the slice-level sparsity exploitation functionality. The ISS pre-fetcher is designed with the FIFOs and its internal ZS detector which judges whether the input slice is zero-bit-stream or not. It fetches only non-ZS (NZS) into FIFOs and increases ZS length (ZSL) if it detects the ZS. On the contrary, cumulative ZSL is extracted when it detects NZS and the ZSL is used as the address of the weight buffer. The NZSs temporally stored in the FIFOs are sequentially used as the input operand of the PE array. ISS has a big potential to improve throughput during the DNN training, and there are two design parameters

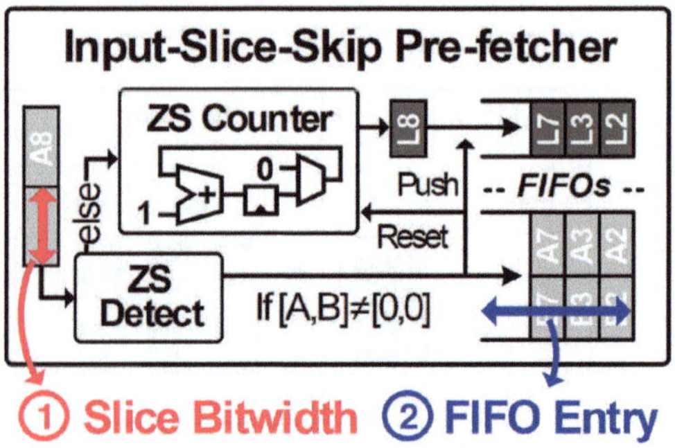

Fig. 6.28 Basic architecture of ISS pre-fetcher

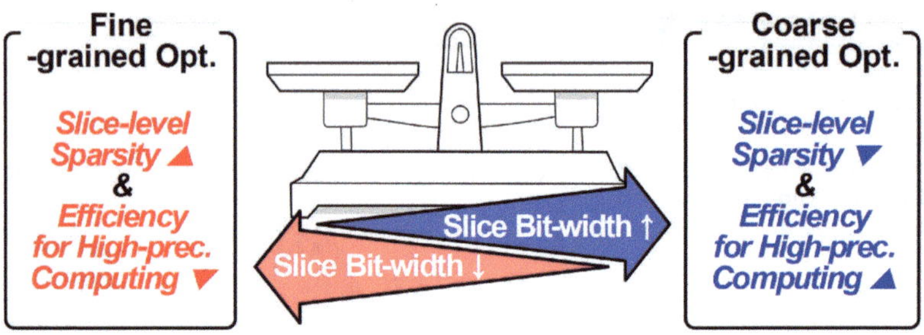

Fig. 6.29 Trade-off according to the slice bit-width

that significantly affect its performance: (1) slice bit-width and (2) the number of FIFO entries.

As described in Fig. 6.29, ISS based core has different pros and cons depending on the slice bit-width. If it uses low slice bit-width, it can finely control the total operation time of the BSSA according to word bit-width. In addition, it induces a high sparsity ratio as described in Figs. 6.30 and 6.31. For the view of sparsity exploitation, lower slice bit-width is more advantageous (Fig. 6.32). However, its computation time can be significantly increased during the high-precision training. Although lower slice bit-width linearly reduces the hardware cost of the basic MAC unit, it increases the computation time quadratically. Figures 6.33 and 6.34 summarize the LAPS based ResNet training results, and this measurement shows that the lower slice bit-width is efficient only when the training task is simple enough. The proposed processor finally adopts 4-bit as the slice bit-width to realize general-purpose training even in difficult tasks such as ImageNet training.

The second factor is the number of FIFO entries. The proposed ISS architecture accumulates NZS into the internal FIFOs and it cannot show ideal performance when the FIFO has a small entry size. That is because it prevents the reading of new input operands from the memory if there is no remaining space to store NZS. However, blindly increasing the number of FIFO entries is not a good solution because it reduces area and energy-efficiency. Figure 6.35 shows simulation results to understand the relationship between FIFO entry and throughput improvement.

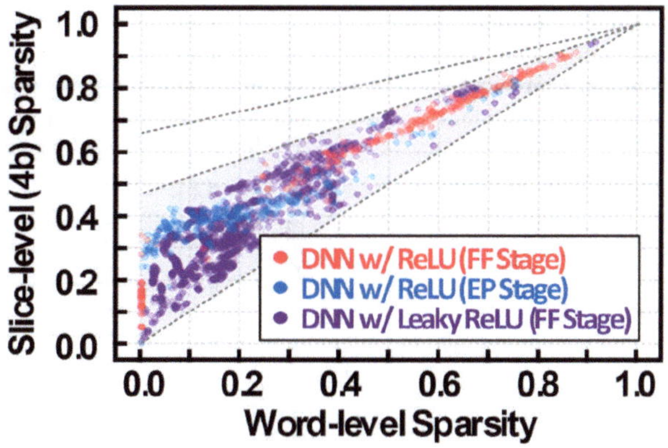

Fig. 6.30 Relationship between word-level sparsity and slice-level (4-bit) sparsity

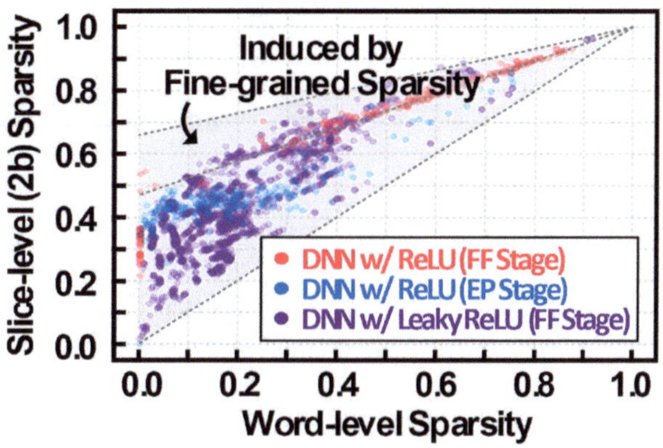

Fig. 6.31 Relationship between word-level sparsity and slice-level (2-bit) sparsity

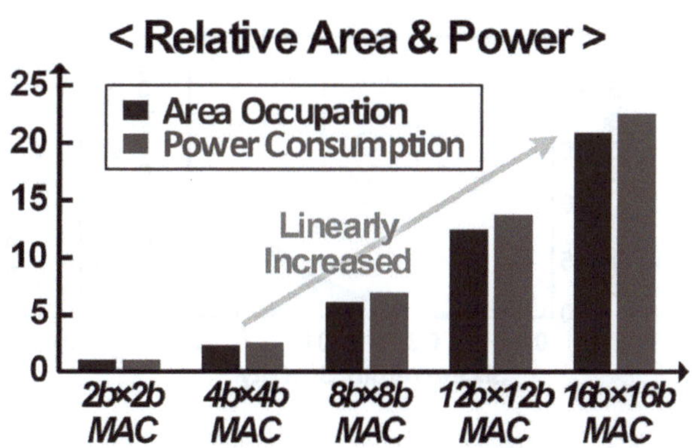

Fig. 6.32 Relative area and power consumption of MAC according to the slice bit-width

The throughput improvement effect of ISS is converged after entry four, so the four-entry FIFO is adopted in this processor.

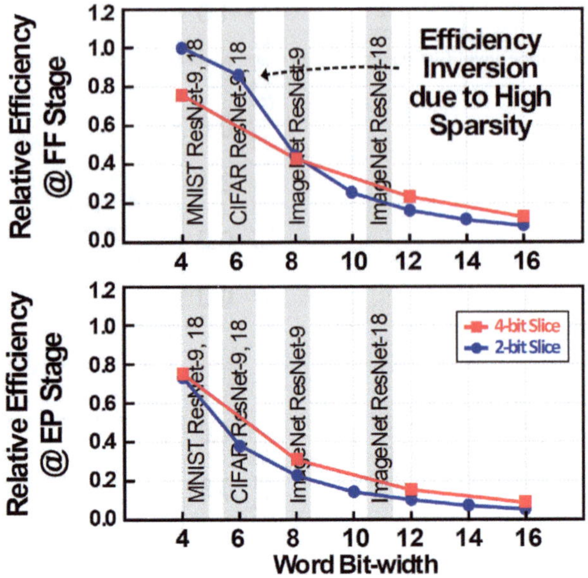

Fig. 6.33 Comparison of energy-efficiency between 2-bit and 4-bit-slices

Fig. 6.34 Average precision determined by LAPS

Network	ResNet-9		ResNet-18	
	FF	EP	FF	EP
MNIST	4.41	4.23	4.13	4.08
CIFAR-10	6.05	5.63	5.03	5.25
CIFAR-100	6.59	6.47	5.60	5.80
ImageNet	7.91	7.89	10.83	11.46

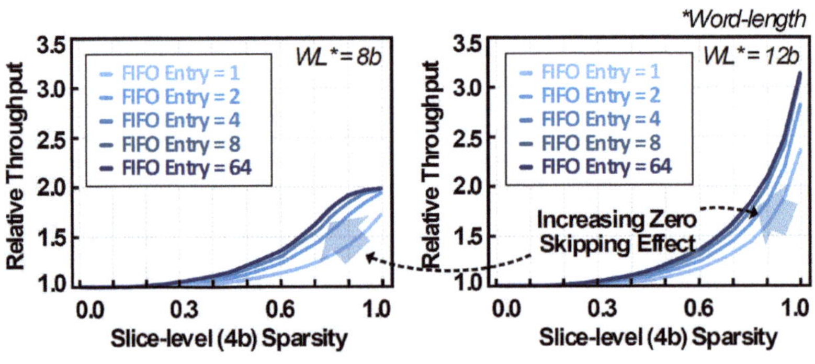

Fig. 6.35 Throughput variation with the number of FIFO entries

Precision-Aware Output Skipping

Unlike FP based computing, FXP based computing induces high multiplication bit-width. On the contrary, the required bit-precision of the next layer is similar to its operands. This characteristic makes negligible accumulations during BSS computing. For example, 12-bit × 12-bit multiplication requires 24-bit for output representation, and it requires a larger bit-width to cover multi-channel accumulation without truncation. Even though the 12-bit × 12-bit convolution induces high accumulation bit-width, the required bit-width for the next layer is as similar as 12-bit. The accumulation results should be truncated to 12-bit and the LSB accumulation information does not affect the final results. Precision-aware output skipping finds out these negligible computations in the LSB accumulation slices and excludes them before it fetches them as the operands. In the abovementioned example, 12-bit × 12-bit convolution is divided into nine accumulation sequences, and three LSB accumulation slices are excluded in the OSS computing. The throughput improvement appears only when the required precision is more than 4-bit because 4-bit × 4-bit convolution consists of only a single accumulation slice.

If the processor directly uses the FP32 based pre-trained model for inference, the accuracy can be degraded due to the OSS. Network fine-tuning with the OSS scheme can compensate for the performance degradation. Additionally, the DNN training performance is not degraded if the OSS is applied from the scratch training. Figure 6.36 shows how the OSS affects the DNN training performance. Since we adopt SR and it adds random numbers to the LSB accumulation results, the importance of the LSB accumulation values is minor and it does not degrades its training performance.

In–Out Slice Skipping Core Architecture

IOSS core architecture can be divided into two major parts. The first part is the IA pre-fetcher which performs input slice skipping (ISS) by excluding ZS. Each PE row has an IA pre-fetcher and a total of four IA pre-fetchers share an input activation memory (IAMEM). The HNPU uses IA pre-fetchers to detect the ZS, and the internal ZS counter generates zero-slice-length (ZSL) which is used as the address of the weight buffer. Only the remaining NZS is stacked in a FIFO and then fed into the PE arrays (PAs) sequentially. When the NZS is fetched from FIFO, it is broadcasted to compute 8 different output channels. Since the different PA has the role of different input channel computing, the accumulation results are transferred to the aggregation module to complete the convolution operation. The ISS, together

Network	ResNet-9				ResNet-18			
Precision	SDFXP8		SDFXP12		SDFXP8		SDFXP12	
OSS	X	O	X	O	X	O	X	O
Train Acc.	98.64	97.96	98.82	98.73	99.96	99.96	99.97	99.97
Test Acc.	73.27	73.36	73.48	73.77	78.18	77.34	78.37	77.48

Fig. 6.36 On-chip learning processor performance comparison table

with bit-slice serial architecture, shows a 7.9 $\times$ higher throughput at 90% slice sparsity in 8-bit training cases.

The other major part is the output slice skipping (OSS) controller which omits useless accumulations based on the precision information of the current and next layers stored in the precision register file. The main role of the OSS controller is generating sequential BSS related instructions. When the OSS controller judges to skip partial accumulations, it skips corresponding instructions before it affects main core computing. It additionally manages accumulation bit-width by using bit-shift operation. Moreover, it modifies the address of the weight buffer and OAMEM simultaneously. With the OSS, the throughputs of 8-bit, 12-bit, and 16-bit convolution are improved by 33%, 50%, and 67%, respectively. As described in Fig. 6.26, combining ISS and OSS, the IOSS architecture compensates for the throughput degradation that occurred in high-bit-precision computing and it increases the normalized throughput and energy-efficiency by 59.9% and 82.9% compared with the GANPU. One of the main reasons for the improvement is that slice-level skipping induces more sparsity compared with word-level skipping (Fig. 6.37). The other reason is that it can find negligible operations appeared in IA and OA and utilize both without any output prediction unit (Fig. 6.38).

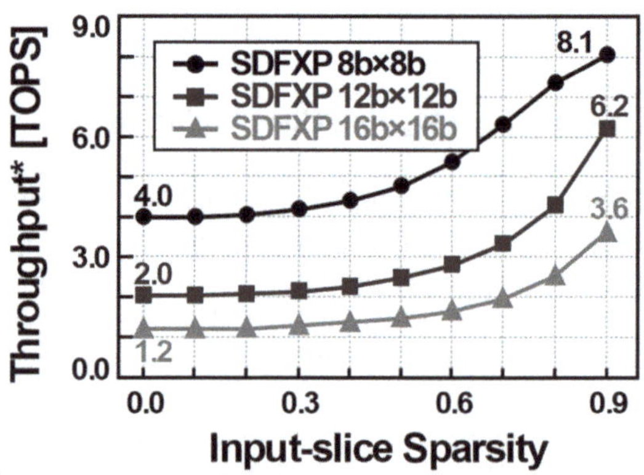

Fig. 6.37 Throughput improvement result according to slice-level sparsity ratio

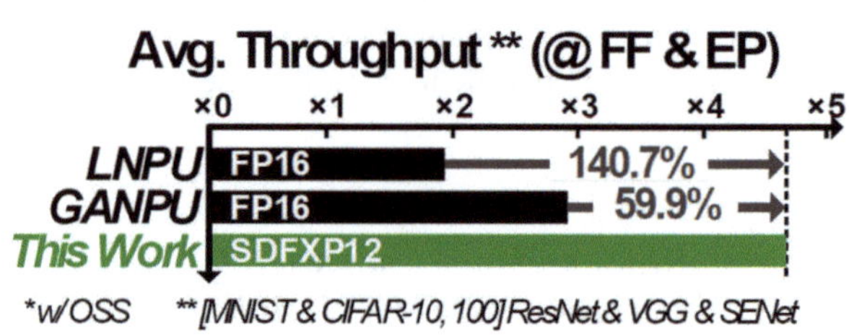

Fig. 6.38 Comparison with LNPU and GANPU was performed in the same 65nm CMOS technology

6.5.3 *Adaptive Bandwidth Reconfigurable Accumulation Network*

The GANPU can support multi-DNN allocation by adopting a reconfigurable accumulation network (RAN) which increases the degree of freedom for the accumulation path. Thanks to the RAN, it can successfully balance computation-bound and memory-bound computing. In spite of its reconfigurable functionality, conventional RAN architecture is based on FP representation and it can support only single precision. It is not suitable for the HNPU because it shows different throughput according to the variable bit-width determined by the LAPS algorithm. The HNPU, therefore, adopts the basic concept of RAN to support multi-DNN allocation but revises it as AB-RAN for variable output bandwidth supporting.

As shown in Fig. 6.39, the main difference of the AB-RAN is bit-sliced adder based accumulation. Bit-sliced accumulation can support variable output bandwidth by dividing or combining multiple adders. The bit-sliced adder of the AB-RAN can process a wide-range output bandwidth with the minimum number of adders and bit-width. Compared with the same number of high-bit-precision adders, the bit-slice adder based accumulation can increase the core utilization by 27.9% even with the bit-width variation from 4-bit to 16-bit. Furthermore, the AB-RAN includes a statistic unit that sequentially calculates the mean and variance of the convolution results. The mean and variance are also calculated by using bit-sliced adders. The BN unit sequentially calculates mean and variance as soon as the main core generates the outputs, so it can compute BN layers without the main core throughput degradation (Figs. 6.40, 6.41, 6.42, 6.43, 6.44, and 6.45).

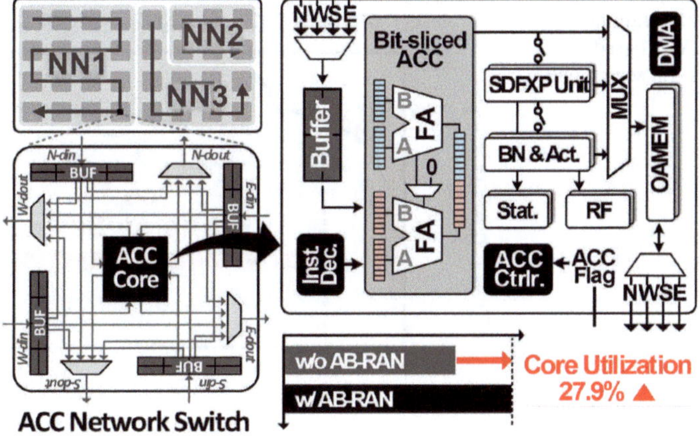

Fig. 6.39 Architecture of bit-sliced adder based reconfigurable accumulation network: AB-RAN

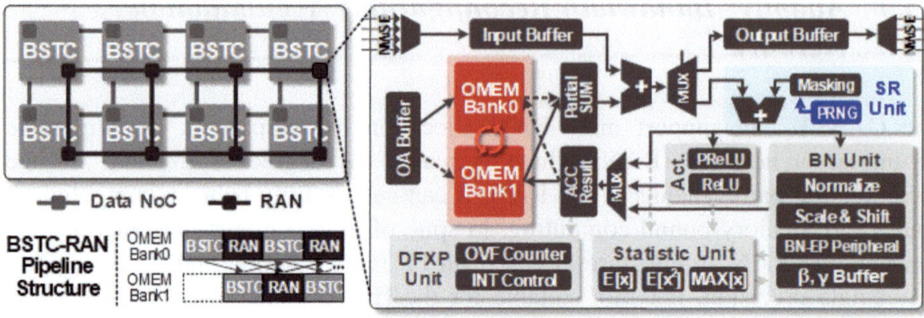

Fig. 6.40 Double-buffered RAN with LT and ST data management

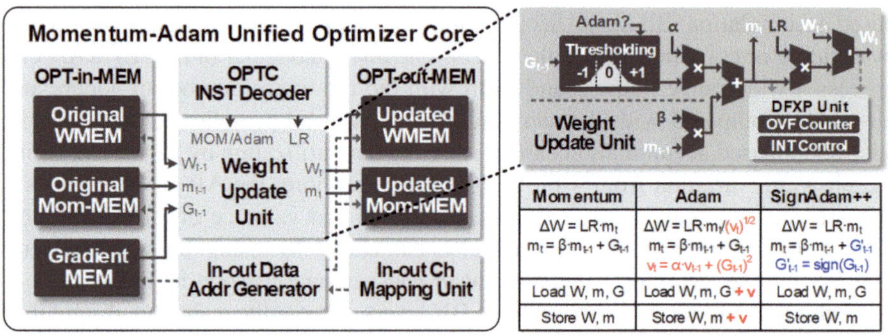

Fig. 6.41 Momentum-ADAM unified OPTC with signAdam++

Fig. 6.42 DNN training curve with LAPS based optimal precision search and IOSS

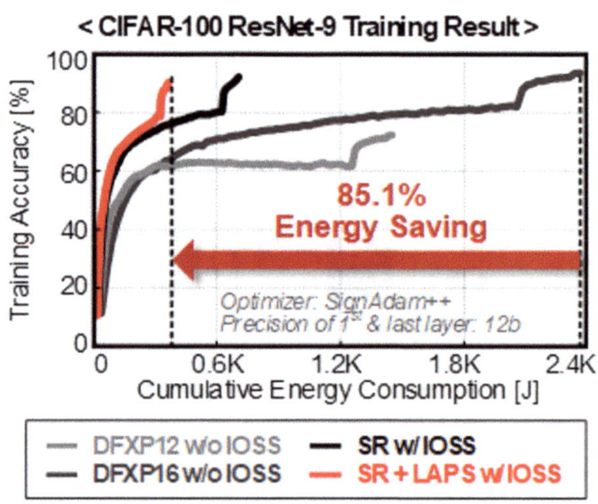

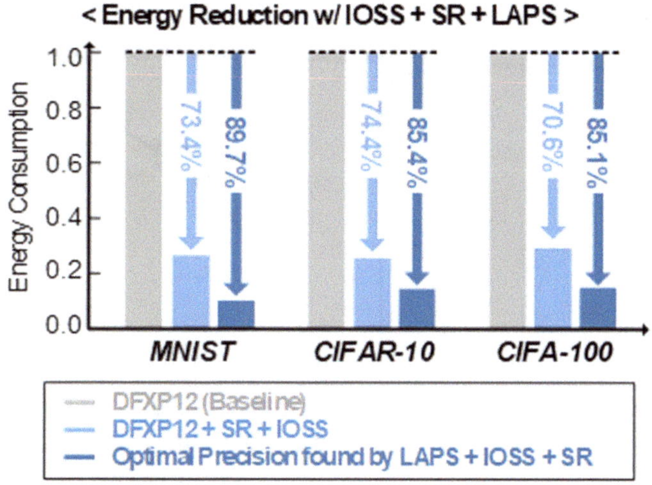

Fig. 6.43 Energy reduction result with LAPS based optimal precision search and IOSS

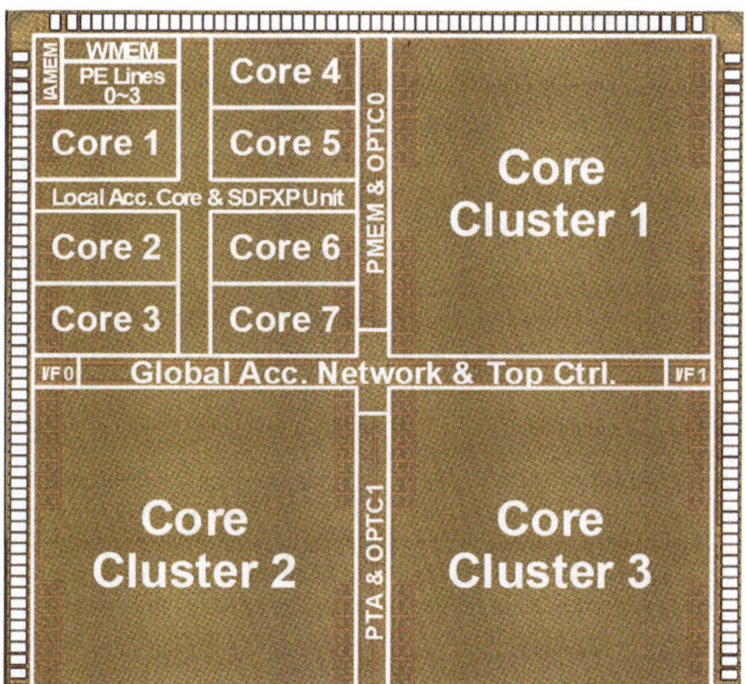

Fig. 6.44 Chip photograph of HNPU

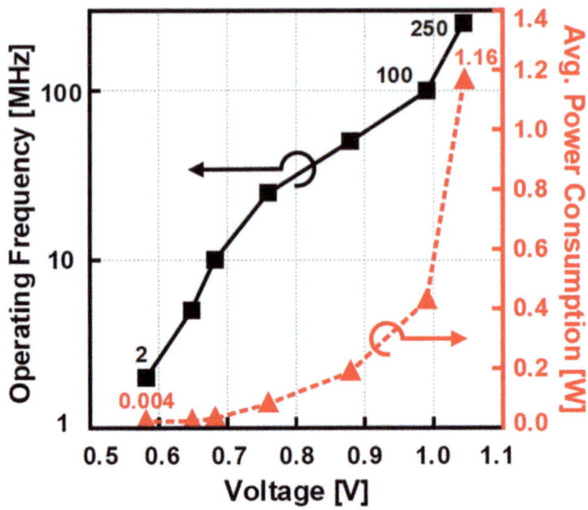

Fig. 6.45 Dynamic voltage–frequency scaling results of HNPU

6.6 Measurement Results

6.6.1 Chip Implementation Results

Figure 6.44 shows the chip implementation result and Fig. 6.46 shows the performance summary. The HNPU is fabricated with the SAMSUNG 28-nm CMOS technology and shows 3.6 × 3.6 mm^2 area occupation. The operating condition is varied from 2 MHz to 250 MHz clock frequency with the 0.58-to-1.04 V supply voltage. Even in the same operating condition, power consumption can also be varied according to input slice sparsity. In the maximum frequency condition, the HNPU shows 500–1162 mW power consumption according to slice-level sparsity. The maximum energy-efficiency appears when the clock frequency is 2 MHz, and the corresponding power consumption appears as 1.95–4.39 mW. Since the ratio of zero-bit-slice is increased in the PE array, it shows lower dynamic power consumption in the higher sparsity ratio.

The throughput of the HNPU can be varied because of the three factors: (1) bit-precision, (2) in-slice sparsity, and (3) OSS. The peak performance always appears at the 4-bit × 4-bit computation, but the corresponding throughput is not varied according to in-slice sparsity and OSS option. On the other hand, the throughput during high-bit-precision computing (>4-bit) is affected by slice-level sparsity and OSS option. When the input slice has 90% sparsity and utilizes OSS, the throughput of the 8- to 16-bit computing shows 2.65-to-4.71 × higher than the baseline. The efficiency of the high-bit-precision training is also affected by its precision and input slice sparsity so that it varies from 1.4 TOPS/W (16-bit without IOSS) to 33.3 TOPS/W (8-bit with IOSS). Although there is no zero-slice skipping during 4-bit computing, the power consumption is reduced due to the zero-gating. The peak

			Specifications			
Technology	Samsung 28nm 1P8M CMOS					
Die Area	3.6mm × 3.6mm (12.96mm^2)					
On-chip SRAM	552 KB					
Op. Condition	0.58V (@ 2MHz) ~ 1.04V (@ 250MHz)					
Data Type	DFXP or SDFXP (4/8/12/16)-bit × (4/8/12/16)-bit					
Area Efficiency	59-to-948 GOPS/mm^2					
	In-slice Sparsity		0%		90%	
Power [mW]	250MHz, 1.04V		1162		500	
	2MHz, 0.58V		4.39		1.95	
			w/o OSS	w/ OSS	w/o OSS	w/ OSS
Peak Performance [TOPS]	250MHz	16b×16b	0.77	1.23	2.27	3.63
		12b×12b	1.37	2.05	4.15	6.22
		8b×8b	3.07	4.10	6.09	8.13
		4b×4b	12.3			
Energy Efficiency [TOPS/W]	250MHz	16b×16b	0.66	1.06	4.53	7.25
		12b×12b	1.18	1.76	8.29	12.4
		8b×8b	2.64	3.53	12.2	16.2
		4b×4b	10.6		24.6	
	2MHz	16b×16b	1.40	2.24	9.29	14.9
		12b×12b	2.49	3.73	17.0	25.5
		8b×8b	5.60	7.46	25.0	33.3
		4b×4b	22.4		50.3	

Fig. 6.46 Performance summary of HNPU

energy-efficiency appears as 50.3 TOPS/W at 90% slice-level sparsity with 4-bit × 4-bit computing.

6.6.2 *CNN Training Benchmark Results*

Many previous papers emphasized just peak performance and energy-efficiency by using high sparsity and low-bit-precision computing conditions. However, it is impractical because most of the DNNs do not be placed in those computing conditions. Specifically, the consideration of low-bit-precision computing should be cautious because it can induce accuracy degradation. For the fair comparison, we composed a CNN training benchmark with three different image classification datasets and widely used networks such as ResNet-9, ResNet-18, VGG-16, and SENet. The measurement results of the first and the last layers are excluded in Figs. 6.47, 6.48, 6.49, 6.50, 6.51, 6.52, and 6.53 because the sparsity pattern of the first and the last layers differs from the other layers.

In the EP stage, the performance of the LNPU [21] is much lower than the GANPU [138] and the HNPU because there is no word-level input sparsity. In contrast, the GANPU can utilize output sparsity, but it still cannot utilize input

			V100 [25]	S. VLSI 19[18]	ISSCC 19 [10]	SSCL 20 [20]	JSSC 20 [19]	ISSCC 20 [11]	S. VLSI 20 [16]	This Work
Skipping Technique			X	X	IA skip	X	X	IA & OA skip	X	IS & OS skip
Process [nm]			12	65	65	65	65	65	14	**28**
Die Area [mm^2]			815	5.76	16.0	16.9	10.24	32.4	9.8	**12.96**
Supply Voltage [V]			-	0.78-1.1	0.78-1.1	0.55-1.0	0.65-1.0	0.70-1.1	0.54-0.80	**0.58-1.04**
Max Frequency [MHz]			1530	200	200	294.4	160	200	1500	**250**
Precision		W	FP 16/32/64	FXP16	FP 8/16	FXP16	DFXP8	FP 8/16	FP 16/32	**SDFXP 4/8/12/16**
		IA	FP 16/32/64	FXP13	FP 8/16	FXP16	DFXP8	FP 8/16	FP 16/32	**SDFXP 4/8/12/16**
Benchmark Average Performance	Avg. Energy Efficiency @ MAX Efficiency Condition[1)] [TOPS/W or TFLOPS/W]	FF	0.42 (FP16)	0.70 (FXP16)	5.11 (FP16)	2.60 (FXP16)	1.03 (DFXP8)	6.93 (FP16)	1.40 (FP16)	**34.06 (SDFXP+LAPS)**
		EP	0.42 (FP16)	0.35 (FXP16)	1.86 (FP16)	2.60 (FXP16)	1.03 (DFXP8)	5.22 (FP16)	1.40 (FP16)	**37.09 (SDFXP+LAPS)**
	Avg. Area Efficiency @ MAX Throughput Condition[1)] [GOPS/mm^2 or GFLOPS/mm^2]	FF	153.4 (FP16)	17.78 (FXP16)	54.70 (FP16)	8.93 (FXP16)	13.10 (DFXP8)	54.01 (FP16)	306.1 (FP16)	**757.61 (SDFXP+LAPS)**
		EP	153.4 (FP16)	8.89 (FXP16)	19.89 (FP16)	8.93 (FXP16)	13.10 (DFXP8)	39.68 (FP16)	306.1 (FP16)	**731.67 (SDFXP+LAPS)**

1) Operating conditions in which the maximum efficiency or throughput of each processor appears

Fig. 6.47 On-chip learning processor performance comparison table (based on benchmark explained in Sect. 6.6.2)

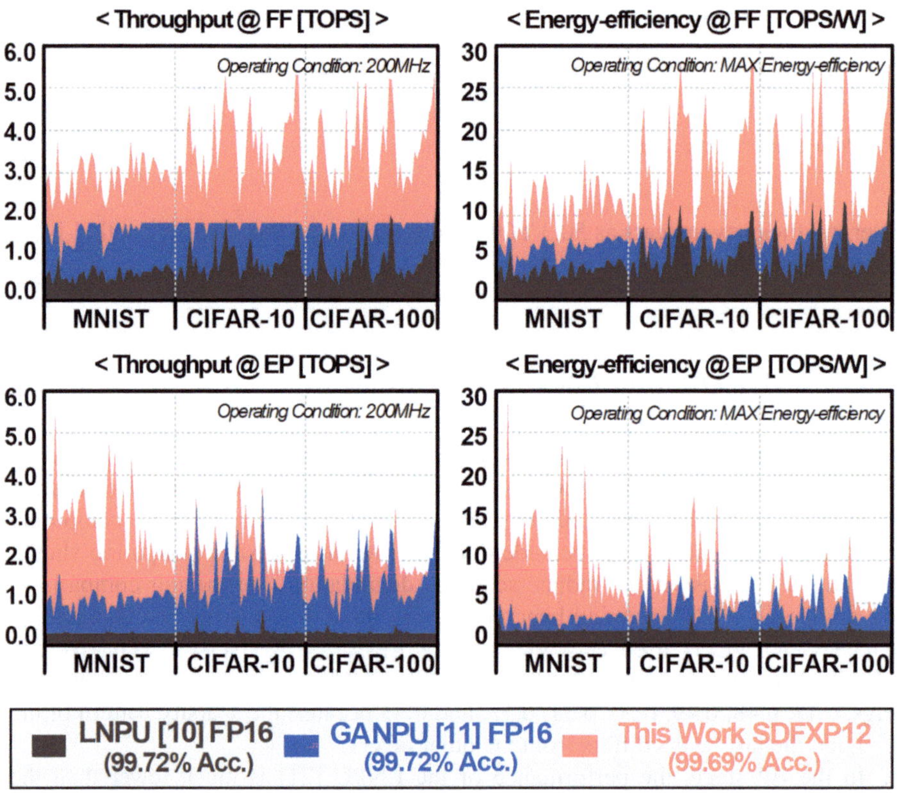

Fig. 6.48 Layer-wise throughput and energy-efficiency comparison result without LAPS. It includes training results of four DNNs (ResNet-9, ResNet-18, VGG-16, and SENet) and three datasets (MNIST, CIFAR-10, and CIFAR-100)

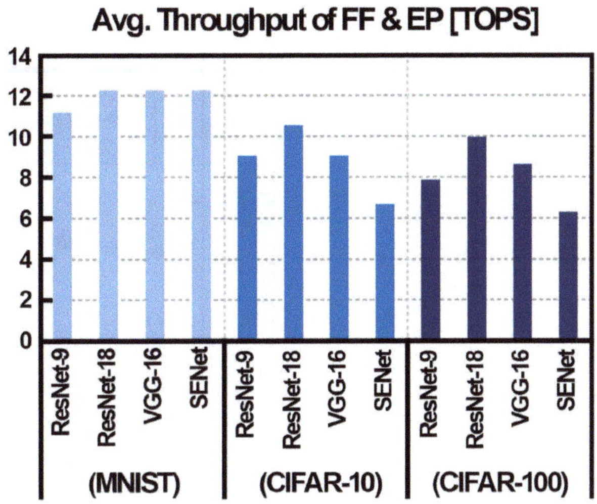

Fig. 6.49 Dataset-wise throughput comparison result with LAPS (measurement result of the last epoch including both the first and last layer results)

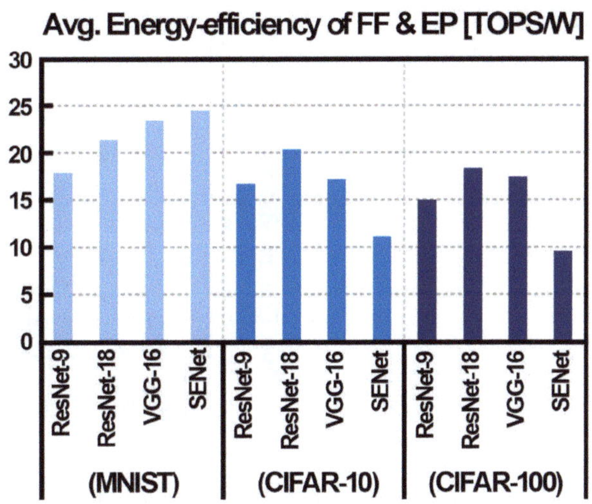

Fig. 6.50 Dataset-wise energy-efficiency comparison result with LAPS (measurement result of the last epoch including both the first and last layer results)

sparsity. Since the HNPU exploits slice-level input sparsity, the HNPU shows higher throughput and energy-efficiency even without word-level output zero-skipping.

In the FF stage, the HNPU still shows the highest performance compared with the other on-chip learning processors. Although the GANPU can utilize dual-sparsity by output zero prediction, the throughput of the main core is restricted because it should wait until the prediction is completed. The HNPU minimizes its energy consumption by excluding output zero prediction and deterministic OSS successfully compensates the efficiency degradation. We summarized the performance improvement results and benchmark comparison results through Figs. 6.48, 6.49, 6.50, 6.51, 6.52, and 6.53. Conventional FXP based training requires 16-bit to achieve high accuracy regardless of network type or dataset. It becomes the main obstacle to energy-efficient DNN training. The HNPU adopts a new number

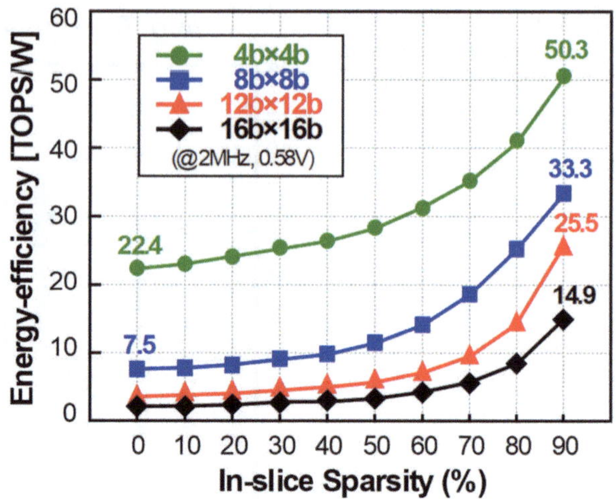

Fig. 6.51 Energy-efficiency according to in-slice sparsity ratio

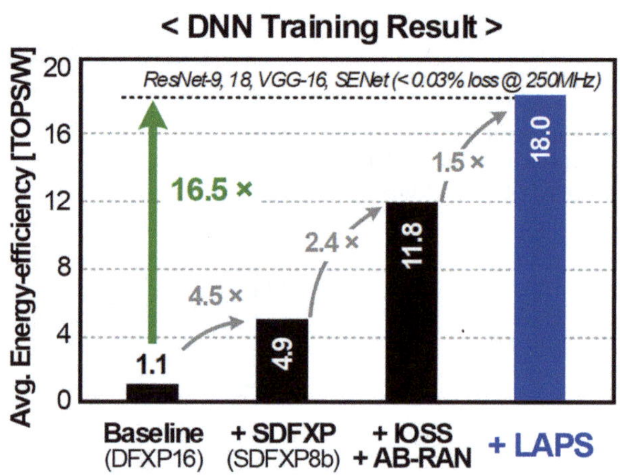

Fig. 6.52 Efficiency improvement result with proposed schemes

representation, SDFXP, and improves the training energy-efficiency by 4.5 ×. In addition, IOSS and AB-RAN further increase its energy-efficiency by 2.4 ×. Finally, the LAPS enhances the energy-efficiency more by 1.5 ×. The LAPS algorithm fixes bit-precision of the first and last layers as 12-bit and determines optimal bit-precision of the other intermediate layers every iteration. Based on the benchmark result, the proposed processor shows at least 5.9 × higher energy-efficiency and 2.5 × higher area-efficiency compared with the previous on-chip learning processors [16, 18, 21, 23, 34, 35, 138].

Figure 6.54 shows the energy-efficiency comparison results during the ImageNet (32×32 resolution) training. Since the LNPU [21] can support FP16/FP8 mixed-precision, we adopt a 0.1/0.9 ratio of FP16/FP8 for ImageNet training. Although GANPU [138] cannot support mixed-precision, it can utilize word-level output sparsity during the EP stage. Moreover, it can support dual-sparsity during the FF stage. The proposed processor, HNPU, shows the highest energy-efficiency

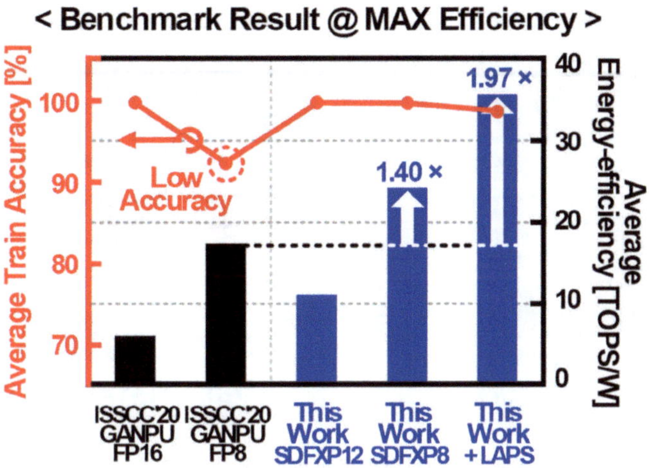

Fig. 6.53 Training accuracy and efficiency comparison with GANPU (based on benchmark explained in Sect. 6.6.2)

Network	ResNet-9			ResNet-18		
Processor	LNPU [10]	GANPU [11]	HNPU	LNPU [10]	GANPU [11]	HNPU
FF Stage	2.41	7.03	14.46	4.27	7.14	15.16
EP Stage	4.68	5.49	5.69	8.20	5.61	6.17
Average	3.54	6.26	10.08	6.24	6.37	10.66

Fig. 6.54 Energy-efficiency (TOPS/W or TFLOPS/W) during the ImageNet (32×32 resolution) training

thanks to the slice-level input sparsity exploitation and layer-wise bit-precision optimization with a new number representation, SDFXP.

As analyzed in the previous work [35], the WG stage can be considered as the computing of a fully connected layer. This characteristic disturbs spatial data reuse that appeared in the convolutional layer so that it induces 33.3% core utilization in the HNPU. Furthermore, the proposed processor adopts high-bit-precision primal weight, so it can utilize ISS but OSS cannot be applied. The throughput improvement through the ISS can also be limited because the main reason for the throughput bottleneck in the WG stage is limited external memory bandwidth.

6.6.3 Online Learning Demonstration Result

In Sect. 6.6.2, the HNPU shows energy-efficient DNN training results. As mentioned in Sect. 6.1, the DNN training can give an opportunity to improve the

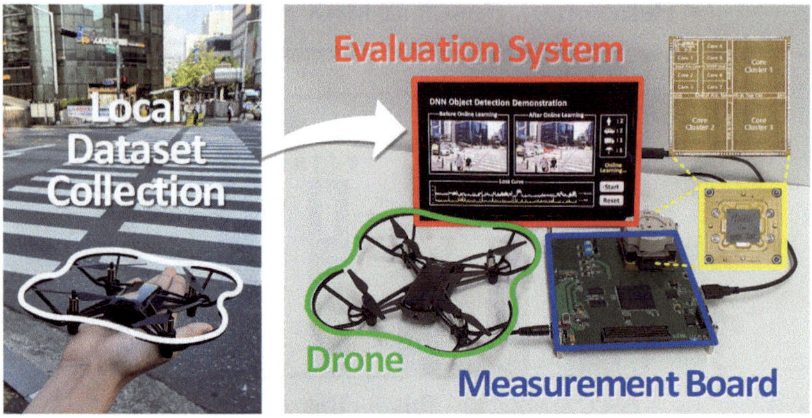

Fig. 6.55 Demonstration system: object detection in drone platform

performance of the mobile-oriented network in practical situations. Online learning can be a solution of accuracy compensation, but the labeling was the main restriction to utilize the training in practical scenarios. TKD [42] suggested self-labeling-based online learning, and it can optimize both the throughput and energy-efficiency during the DNN inference. TKD adopts knowledge distillation which gets labels from the large teacher network, and its inference results are used as the label to train the small student network.

We demonstrated our processor with the TKD based online learning scenarios. The student network is mainly used, but its parameters are updated when the detection accuracy becomes low. In addition, it periodically distills the knowledge of the teacher network to adapt to the environmental changes. The HNPU spends most of the time accelerating Tiny-YOLO-v3 inference but calculates both inferences of the YOLO-v3 and Tiny-YOLO-v3 every five frames. After that, the HNPU updates the parameters of the decoder layers by utilizing YOLO-v3 inference results and repeats this process for 3 iterations before receiving a new frame.

Figure 6.55 shows the TKD based object detection demonstration results using the HNPU. We utilized DDR4 SDRAM as the external memory to store both network parameters and intermediate data that occurred during the computations. The HNPU utilized SDFXP for a low-bit object detection demonstration. The precisions of the network were determined by the LAPS algorithm during the pre-training. Instead, the precision determined during the pre-training is maintained during online learning. Both YOLO-v3 and Tiny-YOLO-v3 finally required 8-bit precision for inference except the first and the last layers. These two layers used 16-bit precision. Furthermore, the precision requirements for EP and WG stages were determined as 16-bit (Fig. 6.56).

In the YOLO-v3 acceleration scenario, it shows the highest object detection accuracy, mean average precision (mAP). However, the HNPU cannot achieve real-time conditions (>30 frames per second, FPS) because of the heavy computation

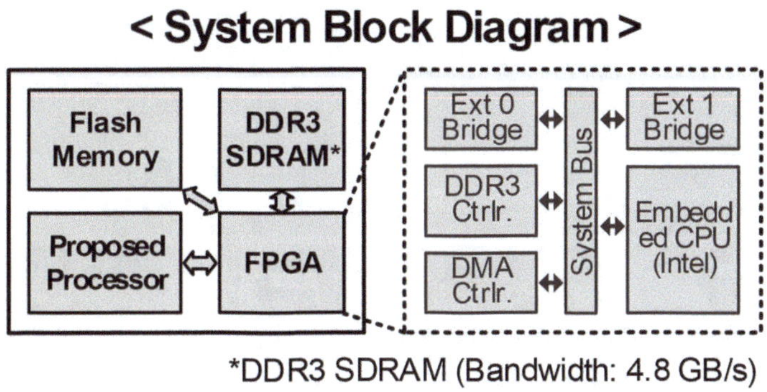

Fig. 6.56 Block diagram of demonstration system

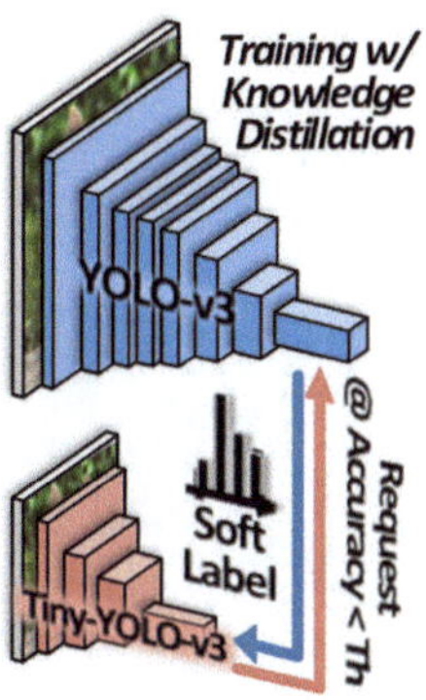

< Custom Dataset Object Detection Result >

	Teacher (YOLO-v3)	Student (Tiny-YOLO-v3)	TKD (Tiny-YOLO-v3)
Accuracy [mAP]	*0.7953*	*0.5935*	*0.7478*
Throughput [frames / sec][1)]	*13.3*	*121.0*	*36.2*
Energy-efficiency [mJ/frame][1),2)]	*37.2*	*3.99*	*13.3*

1) 250MHz, 1.04V, 2) Including DRAM Access

Fig. 6.57 Temporal knowledge distillation (TKD) based object detection scenario

amounts required for YOLO-v3. The HNPU based TKD acceleration shows an average of 36.2 FPS throughput while achieving 0.7479 mAP detection accuracy. Moreover, it can save 64.2% energy consumption compared with the YOLO-v3 acceleration. Thanks to the HNPU based online learning, it successfully achieves high accuracy in object detection applications even in the use of the mobile-oriented network. In addition, the IOSS based bit-slice serial architecture with SDFXP and LAPS algorithm minimizes DNN training overhead and finally accomplishes energy-efficient DNN applications in mobile devices.

In addition to CNN training from the scratch, we demonstrate online on-device training with the proposed processor. Baseline Tinyyolo-v3 was pre-trained on the COCO dataset and shows low object detection quality on our custom datasets such as recent sports broadcasting videos. In the initial 20 frames, on-device training is performed every frame to quickly adapt to the new inference environments. After that, the training frequency is reduced to once every five frames because the training loss becomes low and stable. As shown in Figs. 6.57, 6.58, 6.59, 6.60, and 6.61, when the network is tuned by the proposed processor, the detection performance of

Fig. 6.58 TKD based energy-efficient object detection demonstration result

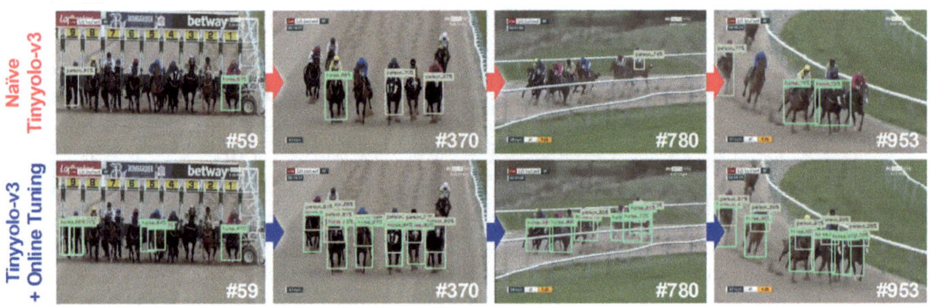

Fig. 6.59 TKD based energy-efficient object detection demonstration result: video sequence

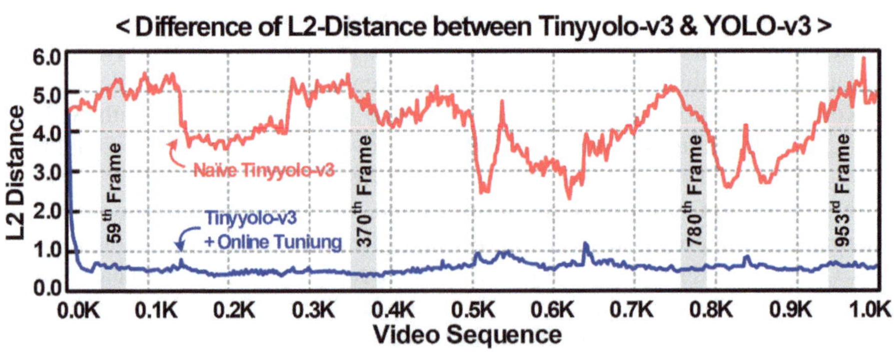

Fig. 6.60 L2 distance between YOLO-V3 and Tinyyolo-v3

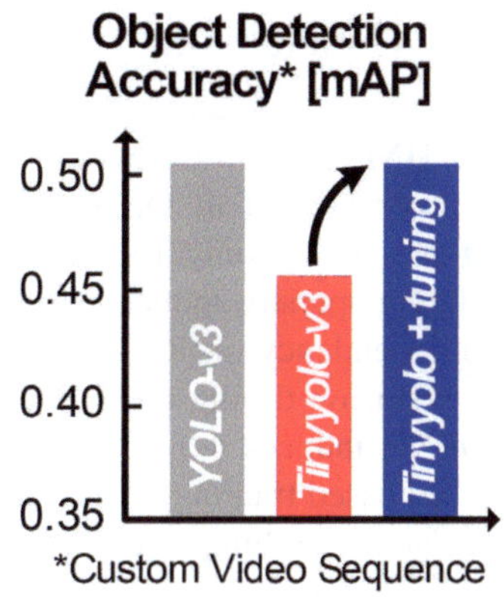

Fig. 6.61 Object detection accuracy in custom dataset

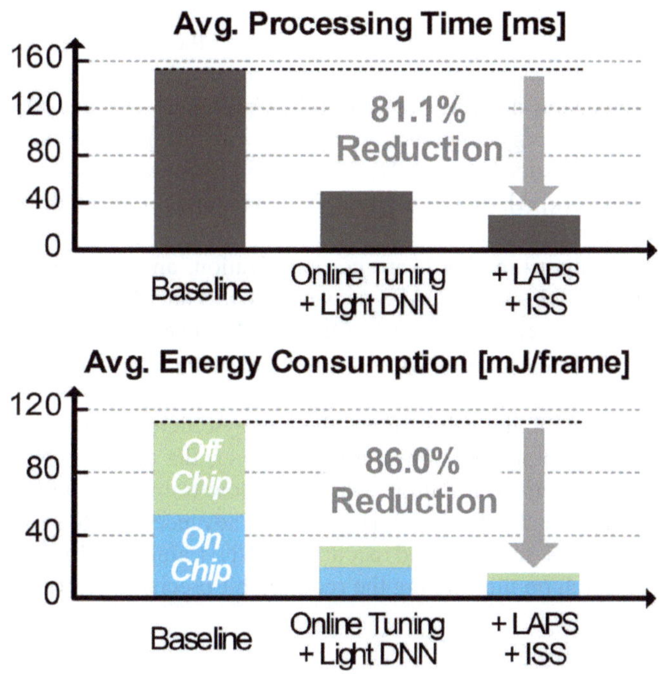

Fig. 6.62 Average processing time and energy consumption of the demonstration system

the network can be compensated even in an unfamiliar situation. During the online tuning, 100-MB capacity DDR3 SDRAM is utilized as the external memory to store network parameters, IAs, errors, and weight gradients. Tinyyolo-v3 online tuning using the proposed processor improves both detection speed and energy-efficiency by 5.3× and 7.1× higher compared to the inference of YOLO-v3 (Fig. 6.62).

6.7 Conclusion

The HNPU, an ultra-low-bit-width (<8-bit) DNN training processor is proposed for energy-efficient DNN training in mobile devices. Conventional DNN training processors only focused on improving throughput and energy-efficiency by utilizing sparsity induced by the ReLU. However, the proposed processor lowers the dependency on ReLU and instead, and it automatically optimizes its required bit-precision during the DNN training. The proposed new number representation, SDFXP, induces 4.5 $\times$ higher energy-efficiency compared with the conventional DFXP based training. When the SDFXP representation is combined with the active precision searching method, LAPS, it can achieve 53.1% higher energy-efficiency in the CNN training benchmark. Not only bit-precision but also slice-level sparsity is utilized to maximize both throughput and efficiency. Slice-level input skipping can exploit 1.38 to 42.4 $\times$ more sparsity compared with the word-level zero-skipping core architectures. Furthermore, it skips useless accumulations according to layer-wise variable precision determined by LAPS. Combining ISS and OSS shows at least 5.9 $\times$ higher energy-efficiency compared with the previous state-of-the-art DNN training processors [16, 18, 21, 23, 34, 35, 138]. At last, the new internal accumulation network, AB-RAN, helps HNPU to maintain its high throughput even in various bit-width conditions.

Thanks to the algorithm–hardware co-optimization, an adaptive DNN training processor with the SDFXP and LAPS as well as IOSS and AB-RAN accomplishes state-of-the-art energy-efficiency, 50.3 TOPS/W. Moreover, the HNPU successfully demonstrates both CNN training from scratch and online learning-based optimization.

The proposed LAPS algorithm shows the energy-efficient DNN training thanks to the automatic optimal precision search. However, the current LAPS algorithm induces lower bit-precision training, so it shows low training accuracy during the ImageNet training. In the future work, we will advance LAPS to support both energy-efficient DNN training and accurate DNN training. Moreover, we will continue the research on minimizing the required bit-precision of primal weight.

Chapter 7
HNPU-V2: An Energy-Efficient DNN Training Processor for Robust Object Detection with Real-World Environmental Adaptation

A DNN training processor with a maximum of 332 TOPS/W is proposed for efficient and robust object detection. The proposed processor is able to support both quantization and pruning-based personalization to make a user-optimized lightweight network. In addition to personalization, it supports real-time adaptation to compensate for accuracy degradation caused by environmental changes or unpredictable situations. It maintains conventional input slice skipping architecture and stochastic rounding-based computing for the efficient acceleration of the DNN training. It further improves efficiency by removing pseudo-RNGs during the stochastic rounding and adding blocks to pruning-aware training. Moreover, it suggests an LT-flag-based reconfigurable accumulation network and enables multi-learning task allocation for low-latency DNN training with the backward unlocking solution. Fabricated in 28-nm technology, the proposed processor demonstrates 46.6 FPS object detection with 0.95 mJ/frame energy consumption that is the state-of-the-art performance compared with the previous processors.

This chapter is based on "A 0.95 mJ/frame DNN Training Processor for Robust Object Detection with Real-World Environmental Adaptation" (IEEE International Conference on Artificial Intelligence Circuits and Systems, [80]).

7.1 Introduction

Object detection is one of the essential applications for the edge devices such as drones or robots. For example, object detection is the basic sensing method for drone navigation or human–robot interaction. Recently, deep neural network (DNN) becomes the mainstream of the object detection algorithm, and its performance continues to be improved by increasing the size of the network. However, smarter DNNs require more parameters, making them difficult to be used on mobile–edge devices that have limited hardware resources. Edge devices inevitably have to use

© The Author(s), under exclusive license to Springer Nature Switzerland AG 2023
D. Han, H.-J. Yoo, *On-Chip Training NPU - Algorithm, Architecture and SoC Design*,
https://doi.org/10.1007/978-3-031-34237-0_7

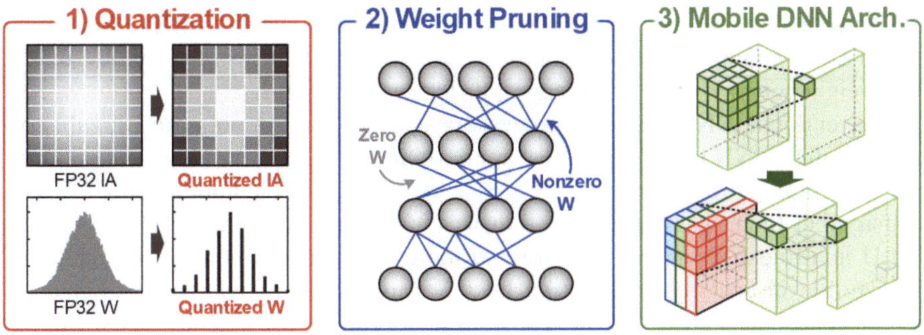

Fig. 7.1 Three techniques for lightweight DNN

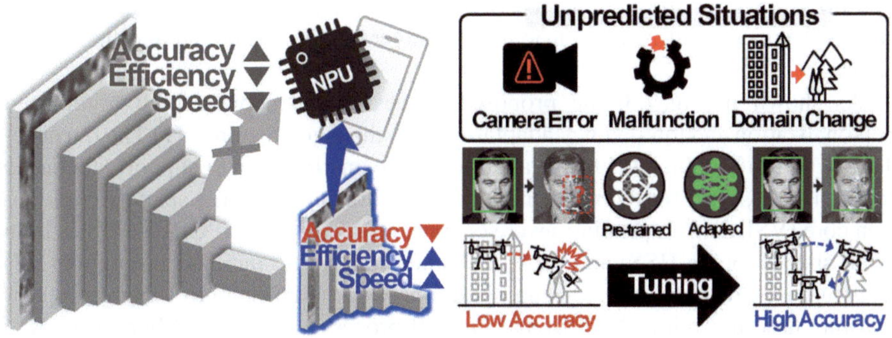

Fig. 7.2 Problems of lightweight DNN and necessity of online DNN tuning for robust object detection

mobile-oriented DNNs that have fewer parameters and computational amounts as shown in Fig. 7.1. One of the trials is the network architecture modification by using simple operations such as 1×1 CONV, but it is still not enough to be realized with the limited power budget.

Another solution is personalization-based network optimization that restricts the functionality of the network only for the user environment. Although the pre-trained DNN targets broad applications, actual usage at the edge device is narrow and biased. It gives an opportunity to reduce the network capacity while maintaining its accuracy on the local dataset. Many DNN training processors [26–29, 34, 123, 138, 141] already adopted personalization-based optimization through the quantization or weight pruning so that they could reduce the computational cost of the DNN dramatically on the edge. Personalization-based lightweight DNN training has shown great success, but it suffers from significant accuracy degradation when it is applied to the new environment. In other words, the lightweight DNN obtained through user customization loses generality due to its low network capacity. In particular, the mobile-oriented DNNs do not work properly in unexpected situations such as camera malfunction as described in Fig. 7.2. Accuracy compensation after an unpredictable accident is important to prevent critical system damage. Real-time online DNN tuning [42] is a promising solution to compensate accuracy of the lightweight network while maintaining its hardware benefits.

Conventional DNN training processors can support training functionality to realize online DNN tuning, but they cannot achieve real-time training. The majority of DNN training processors [26, 27, 138] achieved high throughput by utilizing data sparsity, but their sparsity exploitation was effective only when DNN adopts the ReLU non-linear function. As more and more DNNs adopt non-ReLU functions, the word-level sparsity exploitation becomes useless in popular object detection networks such as YOLO. Moreover, they cannot realize low-latency DNN training because of the backward locking problem [74]. Only one processor [27] tried to solve this problem by adopting a backward unlocking solution (BU), but it was optimized only for fully connected (FC) layers. It shows low reconfigurability so suffers from a core utilization drop in other layer configurations such as the CONV layer.

As shown in Fig. 7.3, the proposed processor can support both high-speed personalization and low-latency adaptation with three key features: (1) The intrinsic-TRNG (iTRNG) generates truly random bit-streams in order to support stochastic rounding (SR, [59]) for required precision reduction. (2) Versatile sparsity exploitation core (VSEC) with pruning-aware channel reordering unit (PCRU) enables both input slice skipping (ISS) and useless channel removal to increase throughput. (3) LT-flag-based reconfigurable accumulation network (RAN) control enables multi-learning task allocation (MLTA) and parallel processing of three different training stages to support BU-based training.

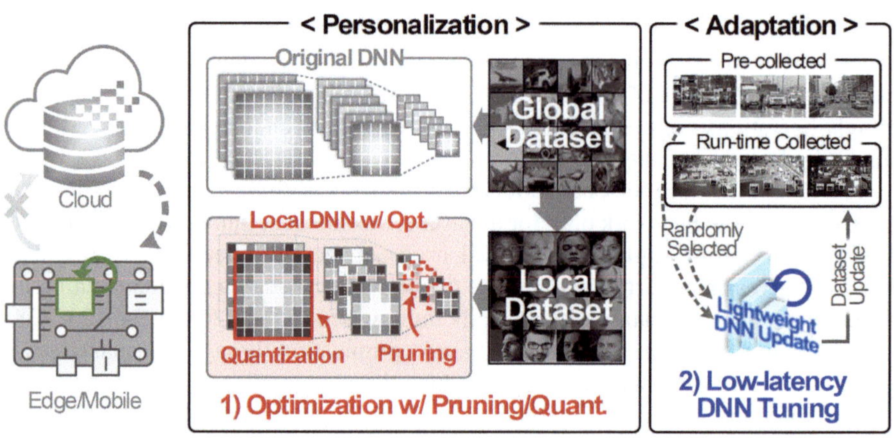

Fig. 7.3 Two types of DNN training supported by the proposed processor

7.2 Overall Architecture

Figure 7.4 describes the overall architecture of the proposed processor. The basic architecture of the proposed processor follows the previous DNN training processor, HNPU [29] for slice-level sparsity exploitation. It includes a total of 32 VSECs, and they are connected with both data NoC and RAN. A single VSEC contains the bit-slice (4b) skip pre-fetcher to omit zero-slice located in the input activation (IA). Since it computes 4b $\times$ 4b at a time, the bit-slice task controller divides computations into multiple μ-instructions to support the high-bit-precision computing. It affects the address generation of IAMEM, weight buffer, and OA buffer. Accumulation switch (ACC SW) placed in every VSEC executes post-processing to computation results. It includes streaming statistics and batch normalization (BN) units. They minimize the throughput degradation caused by external memory access during the BN layer computation. It also includes dynamic fixed-point (DFXP) and iTRNG based SR units for low-bit-precision training. Peripheral training assistant (PTA) consists of a PCRU and an MLTA controller.

Thanks to the ISS, both HNPU and the proposed processor can exploit slice-level sparsity, but the proposed processor can also support pruning-aware training [142]

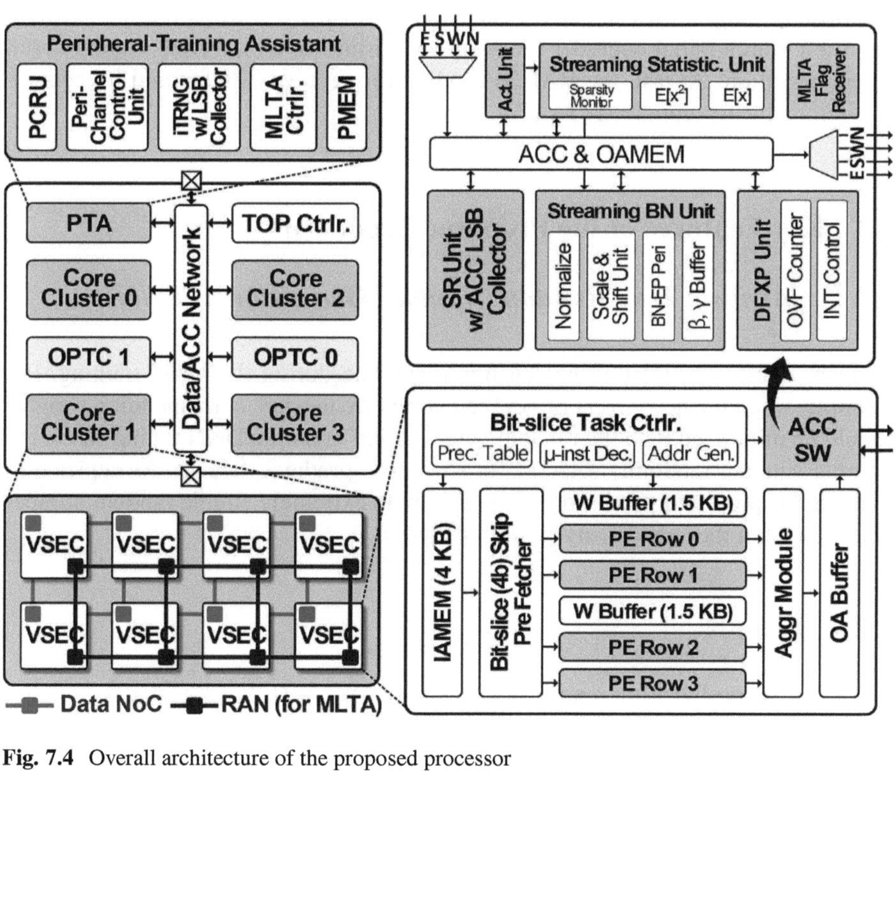

Fig. 7.4 Overall architecture of the proposed processor

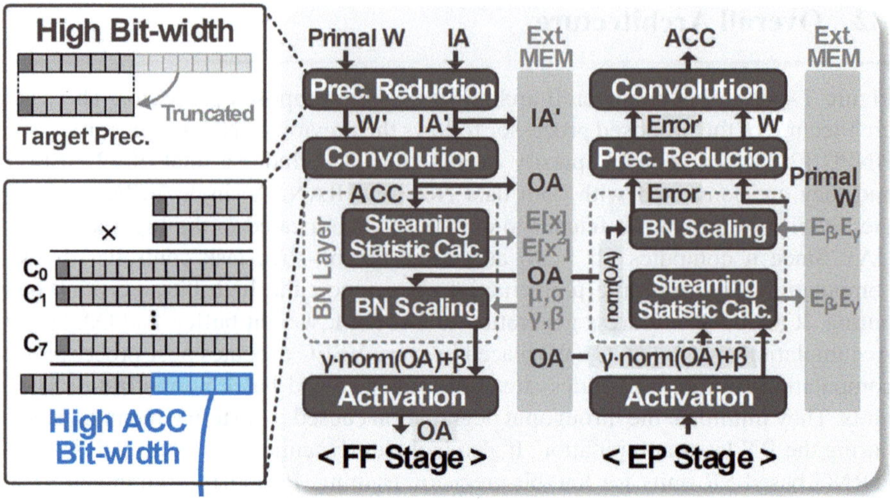

Fig. 7.5 Details of FF-stage and EP-stage computation in HNPU-V2

thanks to the PCRU. Instead, the proposed processor does not include the adaptive precision scaling unit of the HNPU. That is because we verified that the runtime precision search rather degrades the accuracy during the online DNN-tuning-based object detection. At last, the RAN is modified to receive LT-flag, and it becomes programmable to support MLTA (Fig. 7.5).

7.3 Key Features of the Proposed Processor

7.3.1 Intrinsic True Random Number Generator (iTRNG)

Random number generation has an important role during the DNN training. For example, when the new layers or new categories are needed, corresponding weights should be initialized with the random values. Even in online DNN tuning cases, the training data are randomly selected for stochastic gradient descent. As depicted in Fig. 7.6, the proposed processor is designed with extremely simple random number generators, but they show truly random patterns.

iTRNG for DNN

Th DFXP based DNN training, it needs primal weight (PW, [29]) that has higher bit-precision compared with the normal weight used in the feed-forward (FF) and error propagation (EP) stages. In other words, it adopts higher bit-precision only

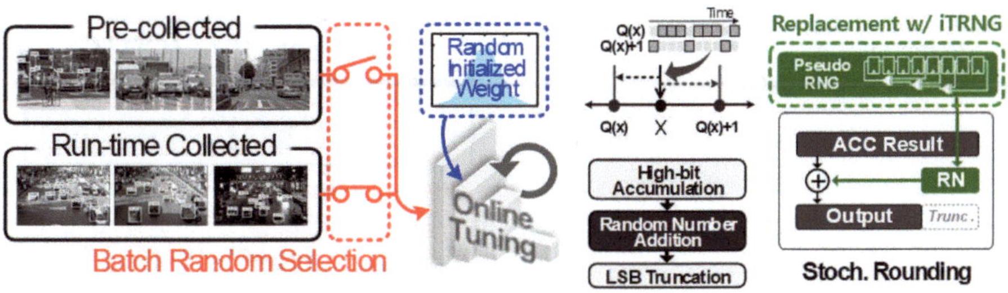

Fig. 7.6 Usage of random number generator in DNN training processor

for the weight-gradient-update (WG) stage. Although the MSB bits of the PW are updated infrequently and show the normal distribution, LSB bits of PW are updated every training iteration, and bit-streams look like truly random patterns. The LSB collector placed in the PTA extracts and cumulates LSB bits of the PW. When 16-bit primal weight is used and the LSB collector extracts LSB 4 bits, it passes all fifteen tasks of the NIST SP-800.22 test [143] with a threshold of P-value > 0.01 (significance level).

iTRNG for Stochastic Rounding

In addition to iTRNG in the PTA, another iTRNG is placed in the ACC SW, and it realizes SR by remaining the LSB bits of the prior ACC results instead of adding random noise. Compared with the conventional SR method designed by the HNPU, it removes all pseudo-RNG (PRNG) circuits from the PEs so that it reduces dynamic power caused by multiple PRNGs. Furthermore, it realizes higher virtual representation resolution compared with the previous linear-feedback-shift-register-based SR. The iTRNG based SR finally achieves 16-bit-level accuracy by using only 8-bit-precision and removes all PRNGs required for SR operation (Figs. 7.7, 7.8, 7.9, 7.10, 7.11, and 7.12).

7.3.2 *Versatile Sparsity Exploitation Core*

Workload Balancing with Sparsity Spreading

As described in Fig. 7.13, the VSEC with the PCRU enables ISS, and it also removes useless channels caused by the narrow distribution or structured pruning. Thanks to the slice-level (4b) sparsity exploitation, the performance improvement is no matter restricted by ReLU. However, the ISS can suffer from the core utilization drop because of the unbalanced workload. The PCRU solves this problem by utilizing

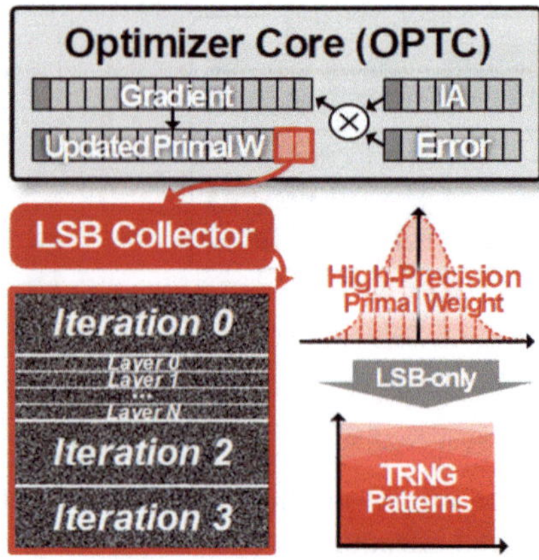

Fig. 7.7 Details of primal weight (PW) based intrinsic-TRNGs (iTRNG)

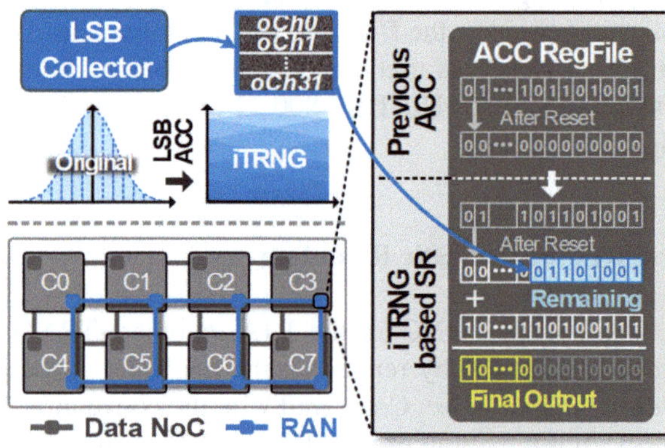

Fig. 7.8 Details of accumulation based intrinsic-TRNGs (iTRNG)

Primal W based iTRNG NIST Pub 800-22 Test*	P-value (After merging 5 epochs)			
	LSB3~0	LSB2~0	LSB1~0	LSB0
Frequency	0.9915	0.8569	0.9819	0.1027
Block Frequency	0.3534	0.6453	0.1071	0.5195
Runs	0.1715	0.7882	0.8931	0.2076
Longest Run	0.4589	0.6490	0.8043	0.8394
Binary Matrix Rank	0.8873	0.4103	0.9985	0.1319
DFT	0.3705	0.3899	0.8406	0.6067
Non-overlapping Template	0.9999	0.9999	0.9999	0.9999
Overlapping Template	0.0383	0.8908	0.6732	0.8166
Universal	0.9992	0.9986	0.9997	0.9995
Linear Complexity	0.2616	0.1387	0.2646	0.8479
Serial	0.8884	0.2642	0.5634	0.1576
Approximate Entropy	0.9264	0.4335	0.7066	0.1566
Cumulative Sums	0.6600	0.9528	0.8181	0.1178
Random Excursions	0.1281	0.4260	0.1272	0.0274
Random Excursions Variant	0.0166	0.0098	0.0056	0.0295

ACC based iTRNG NIST Pub 800-22 Test*	P-value			
	LSB3~0	LSB2~0	LSB1~0	LSB0
Frequency	0.7759	0.7079	0.9924	0.1432
Block Frequency	0.5982	0.8924	0.3000	0.9806
Runs	0.6062	0.0385	0.0455	0.3484
Longest Run	0.8677	0.6089	0.0859	0.4457
Binary Matrix Rank	0.9210	0.9927	0.3583	0.8666
DFT	0.0660	0.2875	0.5735	0.4232
Non-overlapping Template	0.9999	0.9999	0.9999	0.9999
Overlapping Template	0.9953	0.5771	0.3435	0.9537
Universal	0.9997	0.9996	0.9997	0.9992
Linear Complexity	0.5221	0.4601	0.7983	0.2619
Serial	0.5295	0.0893	0.0499	0.5561
Approximate Entropy	0.5275	0.0896	0.0502	0.5542
Cumulative Sums	0.9193	0.5119	0.7242	0.1519
Random Excursions	0.0705	0.0027	0.0610	0.1200
Random Excursions Variant	0.0763	0.1953	0.0188	0.0976

Fig. 7.9 NIST SP-800.22 test results of two iTRNGs

sparsity sorting and spreading. Thanks to the workload balancing with PCRU, it shows a maximum of 21.0% higher throughput.

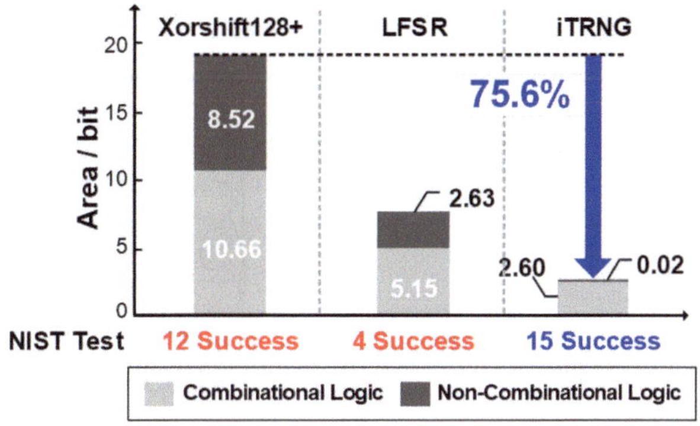

Fig. 7.10 Comparison of conventional PRNGs and iTRNG

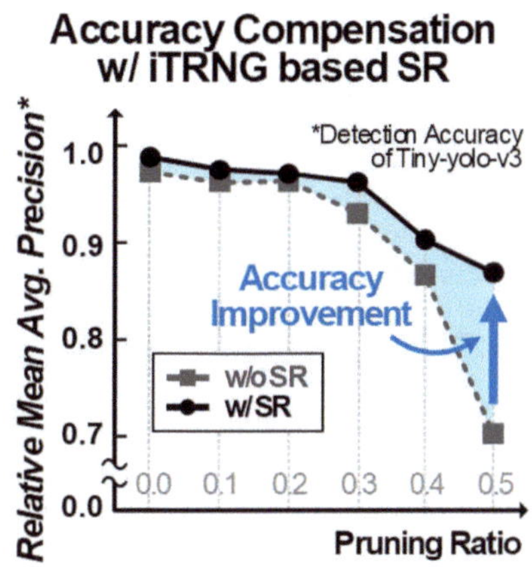

Fig. 7.11 Accuracy compensation result with iTRNG based stochastic rounding

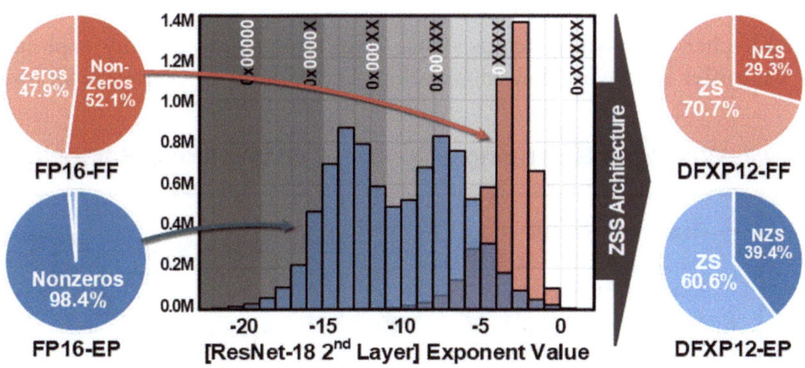

Fig. 7.12 Comparison of word-level and slice-level sparsity appeared during the ResNet-18 training

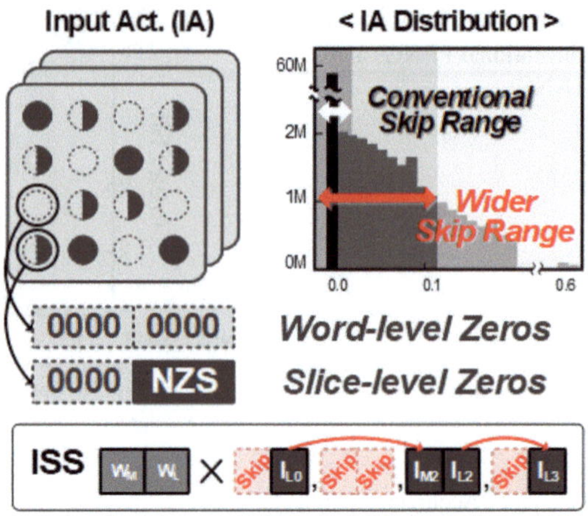

Fig. 7.13 Outline of input slice skipping (ISS)

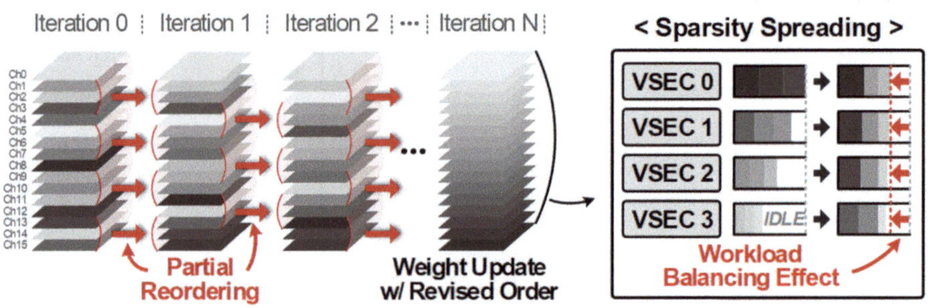

Fig. 7.14 Channel reordering based workload spreading during ISS

Channel Removal with Pruning-Aware Channel Reordering Unit

ISS pre-fetcher successfully realizes slice-level fine-grained sparsity exploitation, but it can suffer from workload unbalancing. The majority of conventional training

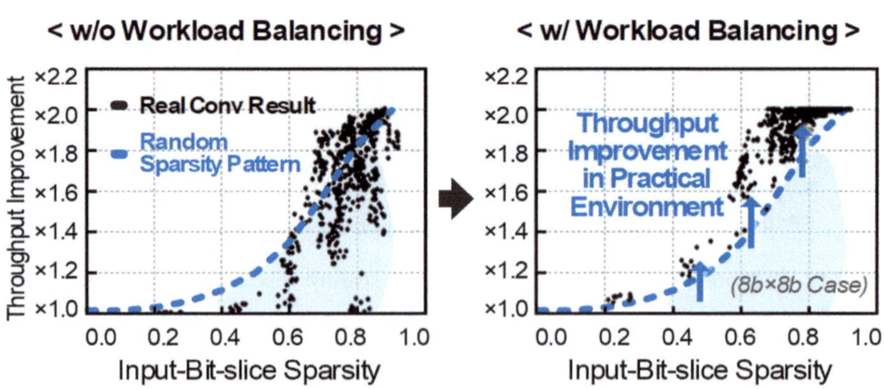

Fig. 7.15 Workload balancing effect with PCRU based ISS

processors [27, 138] overlooked this problem, but it is not negligible on mobile processors because their single core is not big enough to cover large-size batches and channels. One processor [21] tried to solve this problem by designing an internal workload balancer that dynamically switches PE workloads to prevent workload concentration. Even though it achieves high core utilization, it reduces area-efficiency due to the complex routing among the PEs. The overhead of the dynamic workload balancer is further increased in BSSA because the input operand is split into multiple bit-slices, requesting more workload switching. Therefore, a new light workload balancing method is required to maintain high performance even in various sparsity conditions.

The sparsity patterns among the channels are predictable compared with the spatial position or batch. This means that reordering channels by the average sparsity ratio can have strong benefits in balancing workloads. As depicted in Fig. 7.14, the proposed processor adopts both PCRU based gradual channel reordering and sparsity spreading-based task allocation for the workload balancing. The statistic module included in ACC SW counts ZS just before it is moved to output memory. The channel-wise sparsity ratio calculated by the ACC SW is temporally stored in DRAM until the computations of the FF and EP stages are completed.

At the WG stage, the PCRU placed in the optimizer core loads the sparsity ratio of each channel and aligns the channels by adopting sparsity-based partial reordering. It makes 8-input and 8-output channels as a single sorting group and sorts the channels according to the sparsity ratio within the group. The sorting group is interchanged every iteration, and repeating this allows PCRU to gradually sort the entire channels. In the beginning, the sparsity does not differ greatly among the channels, but the difference becomes larger as DNN training goes on. The gradual channel sorting realized by PCRU is sufficient to cope with this sparsity change while minimizing the routing congestion required for sorting. In addition, the weight reordering process does not require additional time because it only needs to write new weight using the modified memory address. After completing the weight update, the proposed processor continues computation of the next iteration with the modified channel order.

PCRU based workload balancing removes dynamic workload balancer and controls only weight update order during the WG stage. It removes additional dynamic power consumption induced by workload switching and minimizes its area occupation thanks to the simple channel sorting method. As shown in Fig. 7.15, ISS shows a maximum of 1.28-to-2.80 $\times$ higher throughput improvement with the help of PCRU and finally achieves 1.4 $\times$ higher energy-efficiency compared with the previous state-of-the-art DNN training processors. Its efficiency is further improved as 1.97 $\times$ when the LAPS unit is combined (Figs. 7.16 and 7.17).

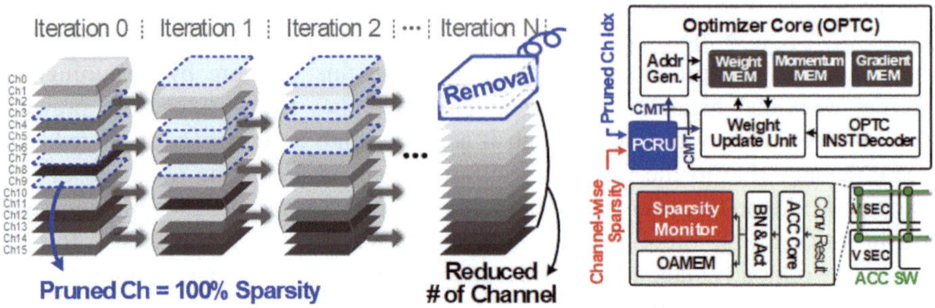

Fig. 7.16 Pruning-aware channel reordering for useless channel removal

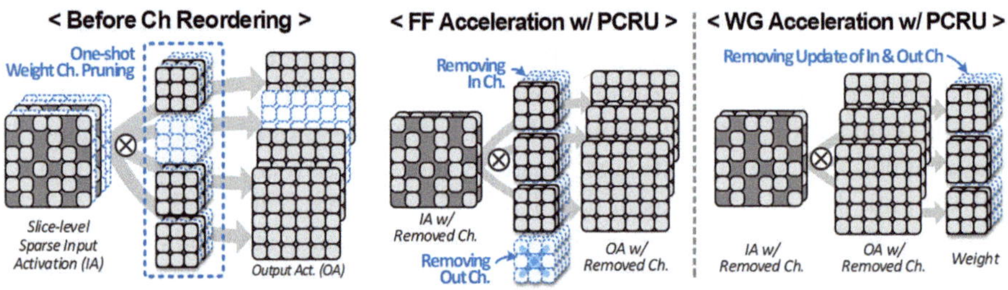

Fig. 7.17 Throughput improvement during the PCRU based DNN training

7.3.3 Multi-learning Task Allocation with LT-Flag-Based Control

Conventional DNN training processor supports multi-DNN allocation through the RAN [138], but it suffers from the backward locking problem. One processor [34] adopts BU, but it can be applied only for fully connected layers because of the low programmability caused by heterogeneous core architecture. As shown in Fig. 7.18, the proposed processor enables not only multi-DNN allocation but also MLTA through the LT-flag-based programmable RAN. The LT-flag includes the information of input–output size, bit-precision, and target training stage. The LT-flag receiver is included in the ACC SW to support dynamic core allocation. It pre-defines data movement among the PMEMs, which are distributed caches of the chip. The LT-flag-based RAN not only controls the accumulation path but also controls data movement among each core to support BUs with various network configurations.

To realize low-latency online DNN tuning, the proposed processor adopts one of the typical BU, direct feedback alignment (DFA, [34, 71, 72, 74]). Unlike the previous processor [34], the proposed processor adopts DFA based training to the CONV layers. As shown in Figs. 7.19 and 7.20, the processor constructs a channel-level pipeline among the three training stages by utilizing DFA based direct

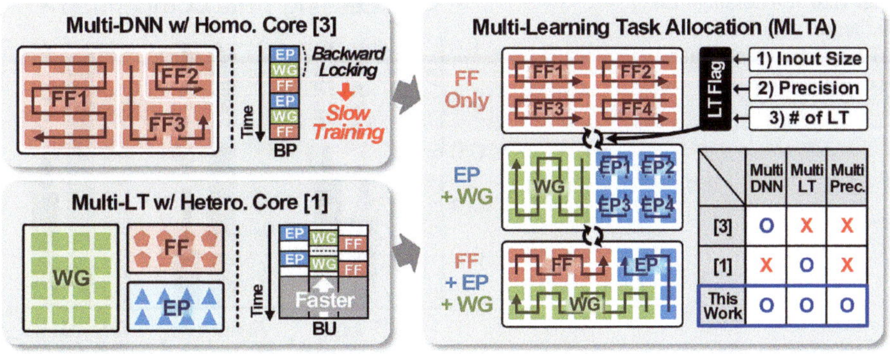

Fig. 7.18 Multi-learning task allocation (MLTA) and comparison with previous processors

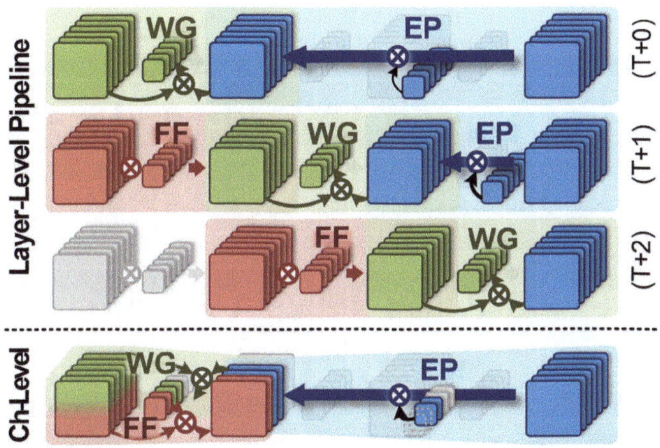

Fig. 7.19 Channel-level pipeline for direct feedback alignment (DFA) based backward unlocking

error propagation. As a result, it shows a 50.3-to-55.8% processing time reduction compared with the back-propagation-based DNN training.

7.4 Measurement Results

7.4.1 Chip Implementation Results

As shown in Fig. 7.21, a 12.96 mm^2 chip is fabricated in 28-nm CMOS technology, and it operates at 10-to-250 MHz core frequency with 0.63-to-1.0 V supply voltage. The proposed processor achieves a maximum of 332 TOPS/W energy-efficiency by utilizing both quantization and weight pruning. Thanks to its lower hardware logic

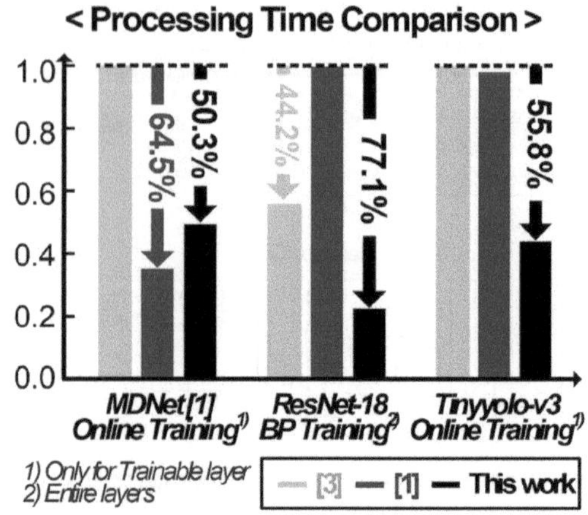

Fig. 7.20 Processing time reduction with the LT-flag-based reconfigurable accumulation network (RAN)

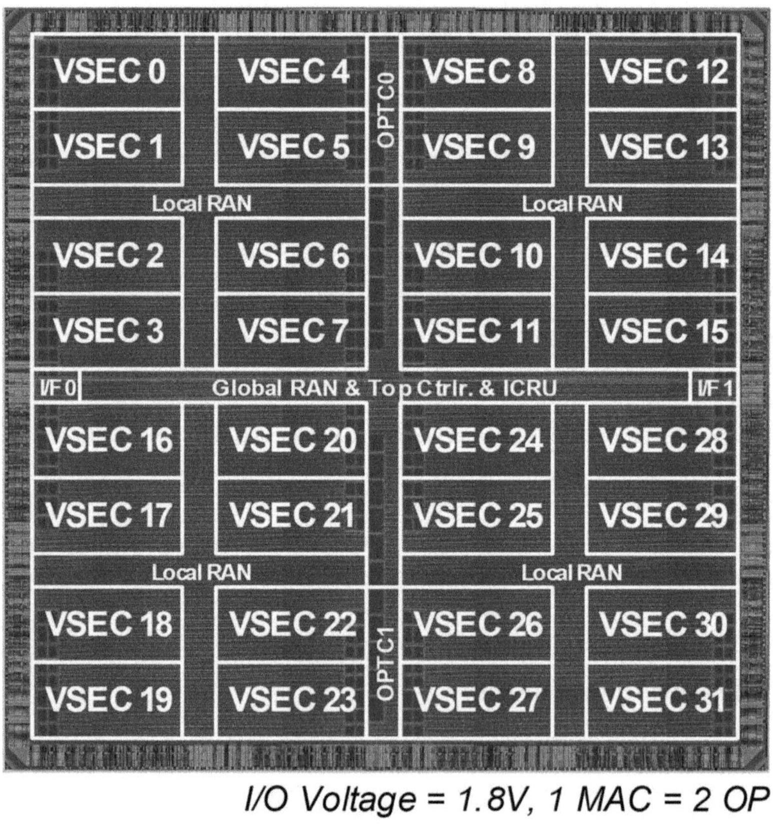

Fig. 7.21 Chip photograph of HNPU-V2

	Specifications			
Technology	Samsung 28nm 1P8M CMOS			
Die Area	3.6mm × 3.6mm (12.96mm^2)			
Supporting BU	BP, DFA [6], ... (Programmable)			
Op. Condition	0.63V (@ 10MHz) ~ 1.0V (@ 250MHz)			
Data Type	DFXP + SR (4/8/12/16)-bit × (4/8/12/16)-bit			
Area Efficiency	59-to-9334 GOPS/mm^2			
	Sparsity (IS, W) [%]	(0,0)	(50,50)	(90,90)
Power [mW]	250MHz, 1.0V	1032	850	616
	10MHz, 0.63V	24.1	19.8	14.6
Energy Efficiency @ 10MHz [TOPS/W]	16b×16b	2.04	6.01	98.1
	8b×8b	6.81	19.9	220
	4b×4b	20.4	49.8	332

Fig. 7.22 Chip performance summary of HNPU-V2

complexity, it also achieves higher efficiency compared to the HNPU [29] regardless of the sparsity condition. As summarized in Fig. 7.23, the proposed processor achieves the best throughput and energy-efficiency during the online DNN-tuning-based object detection scenario. Unlike the conventional DNN training processors, the proposed processor shows high throughput even with the leaky-ReLU thanks to the slice-level sparsity exploitation. It can further improve its efficiency by utilizing weight sparsity induced by pruning (Figs. 7.22 and 7.23).

7.4.2 Object Detection Demonstration Results

The proposed processor is demonstrated with a lightweight object detection network, Tinyyolo-v3. Before it is tested for new video sequences, the network is pre-trained with the COCO dataset and the proposed processor performs both SR-based 8-bit quantization and pruning-aware training. The online DNN tuning is tested in YouTube-Objects (YTO) dataset, and the online DNN tuning is performed only when the detection probability is lower than a threshold. Thanks to the PCRU based pruning-aware acceleration, the energy-efficiency can be improved by 1.6 × higher. The training efficiency is increased up to 3.9 × with the help of SR and ISS. The MLTA over programmable RAN further improves efficiency by 1.7 × higher and enables low-latency training with DFA, significantly lowering the

		[1]	[2]	[3]	[4]	[5]	HNPU-V1 [6]	HNPU-V2
Backward Unlocking		O	X	X	X	X	X	O
Low-precision Training		DFXP	Q-Learning	X	X	FP8-SEB	SDFXP	DFXP+SR[11]
Supporting Sparsity		X	I/O	I/O	I/O/W	X	IS[2)]	IS/OG[3)]/W
Robustness for Non-ReLU		X	X	X	X	X	Zero-slice Skip	Zero-slice Skip
Technology		65nm	28nm	65nm	65nm	40nm	28nm	28nm
MAX Core Frequency		200MHz	268MHz	200MHz	200MHz	180MHz	250MHz	250MHz
Supporting Precision		FXP13/16	FXP2/4/8	FP8/16	FP8/16	FP8	FXP4/8/12/16	FXP4/8/12/16
Throughput [GOPS]	@0% Prune	155	137	1080	610	567	4526	4526
	@20% Prune				763		4526	7072
Area Efficiency[1)] [(GOPS or GFLOPS)/mm^2]	@0% Prune	26.9	24.3	33.3	38.1	77.8	349	349
	@20% Prune				47.7		349	545
Energy Efficiency[1)] [(TOPS or TFLOPS)/W]	@0% Prune	0.62	3.81	1.67	1.44	2.47	4.74	5.05
	@20% Prune				1.79		4.74	7.89

1) Op. condition: MAX throughput, avg. value during Tiny-yolo-v3 online tuning (8-bit, Sparsity by ReLU: 0%, Slice-level Sparsity: 40%)
2) Input Slice (4b), 3) Output Gradient

Fig. 7.23 Chip performance comparison table

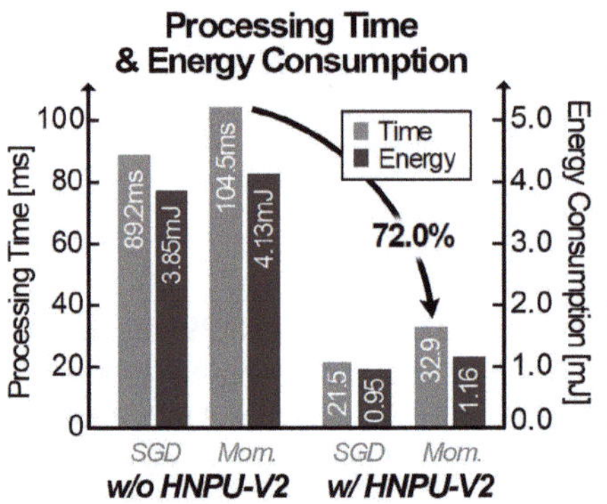

< Object Detection Comparison Table >

	[7]	[1]	[8]	HNPU V1 [6]	HNPU V2
# of Detection	1	1	>1	>1	>1
Backward Unlocking	X	O	X	X	O
Operating Frequency	200 MHz	200 MHz	200 MHz	200 MHz	100 MHz
Framerate (FPS)	30.4	34.4	25.2	26.7	46.6
Energy Efficiency [mJ/frame]	4.14	4.88	2.13	1.68	0.95

Fig. 7.24 (1) Processing time and energy consumption of the proposed processor. (2) Object detection performance comparison table

required operating frequency. The proposed processor finally consumes an average of 0.95 mJ/frame that is the lowest energy consumption compared with the state-of-the-art processors as explained in Fig. 7.24. It also demonstrates the fastest object detection with 46.6 FPS compared to existing processors.

As described in Fig. 7.25, the proposed processor is able to reduce both processing time and energy consumption while maintaining high detection accuracy through online DNN tuning. It shows 15.6%p higher detection accuracy compared with the DNN without tuning case. The proposed processor is also examined with the broken lens environment, and the results are shown in Fig. 7.33. User-customized lightweight DNN shows a low detection rate because it loses detection generality during personalization. Whenever the lens is broken, the proposed processor recognizes low detection probability and begins online DNN tuning. We verified that it can compensate for accuracy degradation caused by an unpredictable

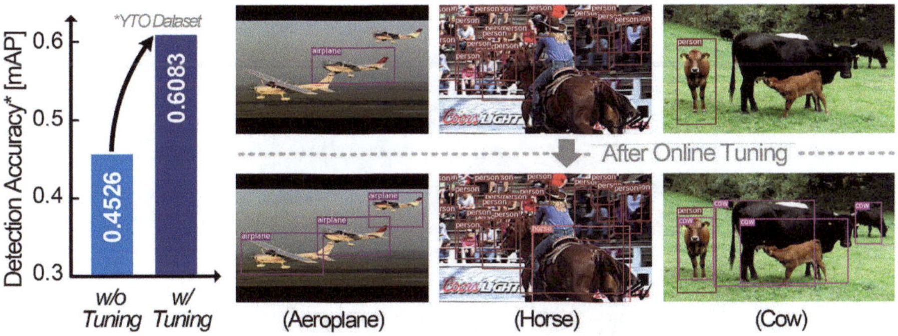

Fig. 7.25 Object detection accuracy measured from the YouTube-Object (YTO) dataset

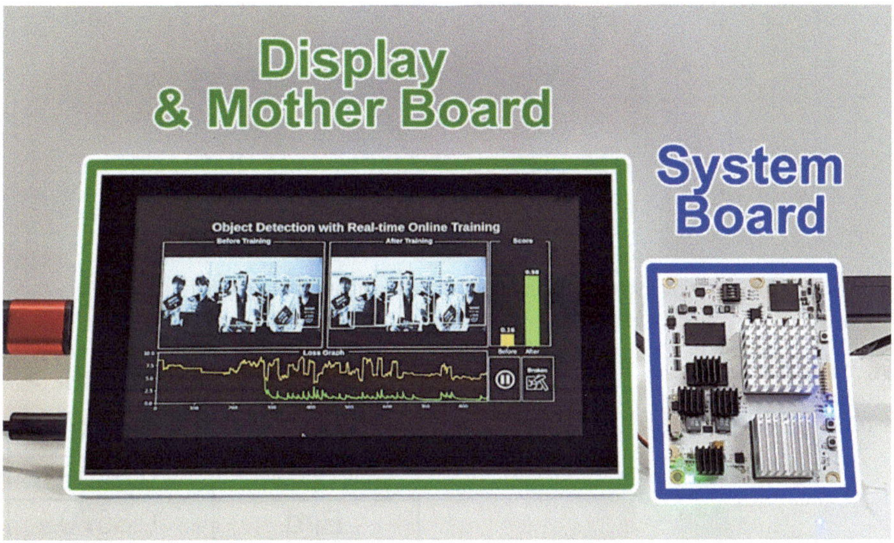

Fig. 7.26 Demonstration system example: HNPU-V2 system board and display

accident such as a camera malfunction. To sum up, the proposed processor recovers its detection accuracy and realizes robust object detection against environmental changes or incidents (Figs. 7.26, 7.27, 7.28, 7.29, 7.30, 7.31, 7.32, and 7.33).

7.5 Conclusion

An energy-efficient and low-latency DNN training processor is proposed for robust object detection with real-world environmental adaptation. The iTRNG based SR enables 8-bit quantization and increases efficiency by 2.2 × higher. A combination

Fig. 7.27 Demonstration system example: Mobile autonomous driving robot

Fig. 7.28 Detailed object detection accuracy in the YTO dataset

< YTO Detection Result >

Class	mAP
Aeroplane	0.6003
Bird	0.8570
Boat	0.5353
Cat	0.5373
Cow	0.5616
Car	0.5911
Train	0.6641
Horse	0.5144
Dog	0.7228
Motorbike	0.5069
Average	**0.6083**

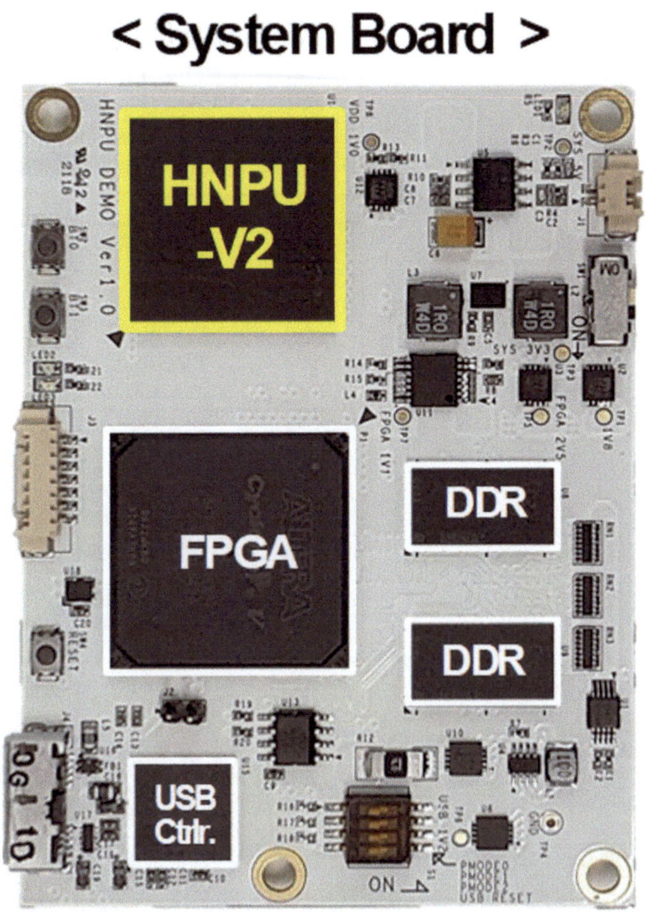

Fig. 7.29 Details of HNPU-V2 system board

of ISS and PCRU exploits sparsity even with the non-ReLU function and achieves 78.4% higher efficiency. When these features are combined with MLTA based BU supporting, it finally shows 6.6 × higher efficiency and 4.2 × higher throughput compared with the baseline. In conclusion, the proposed processor successfully demonstrates the object detection application with the state-of-the-art energy-efficiency, 0.95 mJ/frame, while maintaining 46.6 FPS real-time operation.

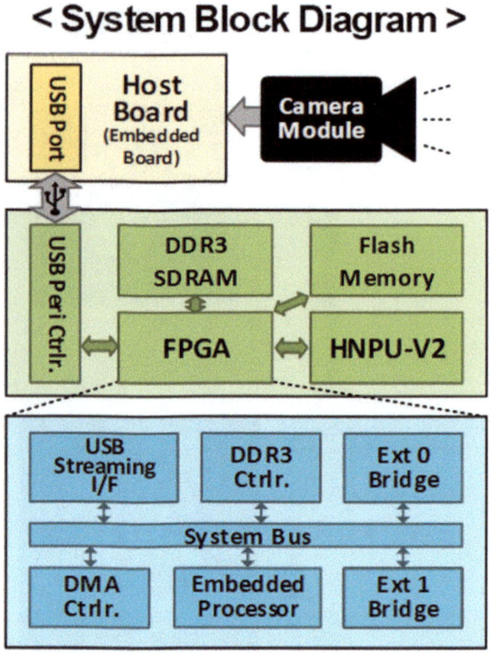

Fig. 7.30 Block diagram of HNPU-V2 system board

Fig. 7.31 Object detection results before and after online DNN training in the YTO dataset

Fig. 7.32 Object detection results before and after online DNN training in the custom dataset (collected video clip)

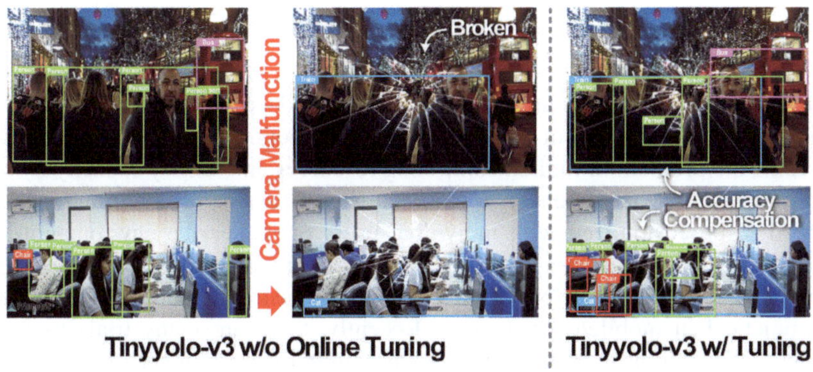

Fig. 7.33 Improved object detection performance in custom dataset (even with the camera malfunction)

Chapter 8
An Overview of Energy-Efficient DNN Training Processors

Many edge/mobile devices are now able to utilize deep neural networks (DNNs) thanks to the development of mobile DNN accelerators. Mobile DNN accelerators overcame the problem of limited computing resources and battery capacity by realizing energy-efficient inference. However, it shows passive behavior and it makes AI hard to be customized with individual users or its service environment. The importance of on-chip training is rising more and more to provide active interaction. Despite its advantages, DNN training has more constraints than inference and it is impractical in mobile/edge devices. Recently, there are many trials to realize mobile DNN training, and this chapter will summarize a number of prior works. Firstly, it arranges the new challenges of the DNN accelerator induced by training functionality and summarizes new hardware features related to the challenges. Secondly, it discusses algorithm–hardware co-optimization methods and explains why it becomes mainstream in training processor research. Thirdly, it compares the main differences between the conventional inference accelerators and recent training processors. Finally, conclusion is made by proposing the future directions of the DNN training processor at the micro-AI systems.

8.1 Solution of Transpose-Read During Error Propagation

As summarized in Fig. 8.1, there are three categories of transpose-read solutions: (1) software-level, (2) architecture-level, and (3) circuit-level solutions.

© The Author(s), under exclusive license to Springer Nature Switzerland AG 2023

D. Han, H.-J. Yoo, *On-Chip Training NPU - Algorithm, Architecture and SoC Design*,
https://doi.org/10.1007/978-3-031-34237-0_8

Level	Paper	Features	Pros.	Cons.	Related Figure
SW or Algorithm Level	[43]	Feedback Alignment [44]	No WT reading	Lossy training + Extra BW	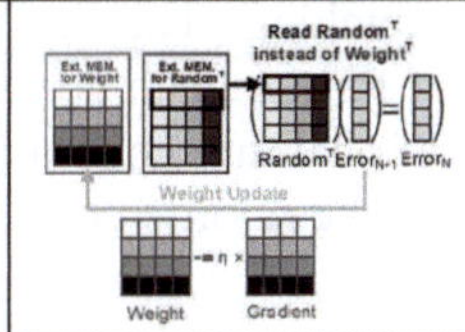
	[22] [29]	Channel Re-arrangement	Off-the-shelf SRAM	Need time to re-arrange weights	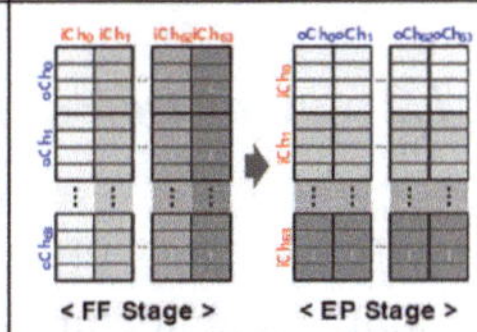
	[23]	Diagonal Storage Pattern	Off-the-shelf SRAM	Need bit-rotator	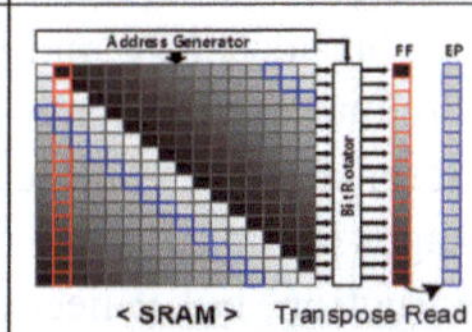
Arch Level.	[20]	Transposable PE Array	Off-the-shelf SRAM	Efficiency drop of PE array	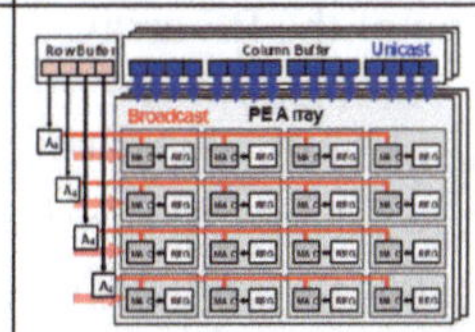
	[30]	Sparse Weight Transposer	Fast Decoding with Sparse Weight	Need Additional Register Array	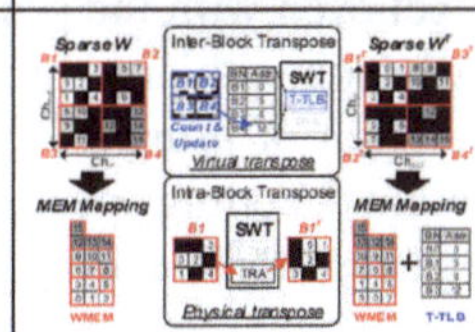
SRAM Level	[5] [25]	Transpose SRAM	Simple PE array arch.	Low data density (Double BL/WL or > 6T cell)	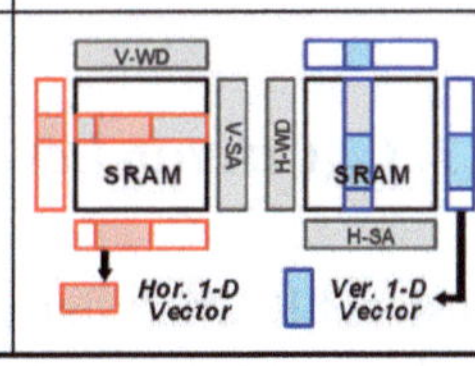

Fig. 8.1 Summary of transposed reading schemes

8.1.1 Software-Level Solution

The authors in [43] tried to avoid transposed weight by using a different EP method called feedback alignment [44]. This algorithm is the same as BP, but the backward weight is substituted with the binary random matrix. However, since this algorithm modification may degrade the training accuracy significantly, it can only be applied to limited DNN applications.

Most of the DNN processors [22, 23, 29], instead, adopted software-level data pre-fetching. They re-arranged the data stored in weight SRAM before it was used for the main convolution operation. This method is straightforward but needs additional time to re-arrange the weight memory. Therefore, it can be efficient only when the re-ordered weights are reused repeatedly.

8.1.2 Architecture-Level Solution

The transposable PE array suggested by [20] releases the burden of the software by simple architectural modification of the PE array. Transposable PE array utilized both broadcasting and unicasting data flow to perform the main computation such as matrix multiplication or convolution. It utilized input feature reuse during the inference with a single image, but it adopted weight reuse by aggregating multiple images in the EP stage. This method can convert its computing type by using simple instruction, but it needs a local buffer to each MAC unit resulting in an efficiency drop of the PE array.

Another paper [30] added a weight transpose-reading unit instead of modifying the main PE array. This additional unit generated the transposed weights before they were fetched to the main PE array. Since the transposable PE array [20] assumed regular patterns of rectangular-shaped weight, there is no throughput improvement effect even with the pruned weight. Sparsity-aware weight transposer suggested by [30] adopted hierarchical transpose-read and reduced memory access by excluding the fetch of pruned weights. Even though the weight transposer unit showed fast weight decoding speed, it can rather induce low area-efficiency due to the additional large register file array.

8.1.3 Circuit-Level Solution

Custom SRAM design [5, 25] is also a good solution which can support both normal read and transpose-read. It makes both main core architecture and software much simpler, but the memory density of the SRAM may be degraded because the SRAM cell architecture should be modified by adding more MOSFETs or bit/word-line.

8.2 Sparsity Exploitation During DNN Training

Section 8.1 explained hardware design methods to accelerate EP with transpose-reading of weight, but it needs further optimization to realize high-speed DNN training at the micro-AI systems. It needs additional hardware features to maximize its throughput even with the limited resources. Sparsity exploitation is the key

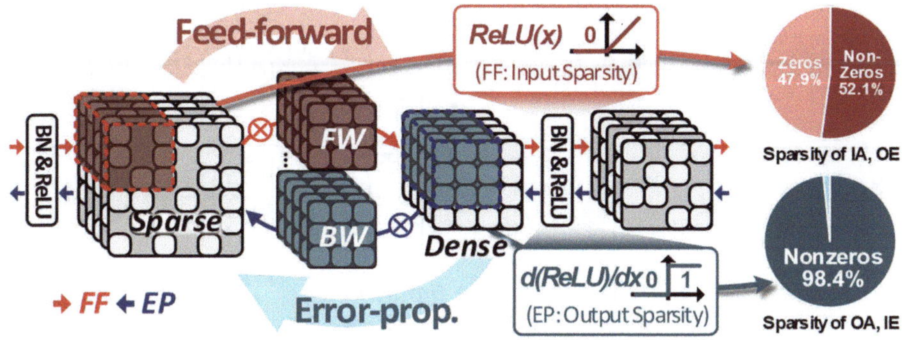

Fig. 8.2 Sparsity appears during the feed-forward (FF) and error propagation (EP) stages

feature to increase throughput during the on-chip training and it is summarized in Fig. 8.4.

8.2.1 Conventional Input/Weight Zero Skipping

Most of the DNNs utilize the ReLU activation function, and it makes the negative values of IA become zeros. Weight can also have many zero values when the pruning is applied. Zero operands do not affect the final output and many inference engines [33, 45, 46] tried to speed up their operations by skipping zeros in the IA or weight. Input and weight skipping techniques are still effective for DNN training, but it shows limited performance improvement during the EP and WG stages. At the EP stage, input sparsity disappears due to the normalization layer (Fig. 8.2). On the other hand, weight sparsity disappears at the WG stage (Fig. 8.3) because the gradient is calculated by the convolution between IA and OE. DNN training processor needs a different sparsity exploitation method compared with inference processor to utilize the advantages of the zeros that appear during the training.

8.2.2 ReLU-Aware Output Zero Skipping During the EP

Input sparsity cannot be utilized for the acceleration of the EP stage. Instead, the derivative of ReLU activation, dReLU(x)/dx, transfers only positive values to the prior layer. It induces output sparsity as shown in Fig. 8.2. Recent DNN training processors [27, 138] exploit both input and output sparsity to improve their throughput in both FF and EP stages.

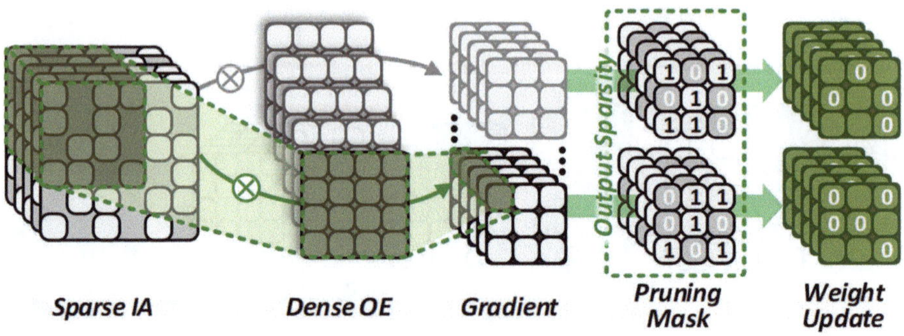

Fig. 8.3 Sparsity appears during the weight-gradient (WG) stage

8.2.3 Pruning-Aware Output Zero Skipping during the WG

Weight sparsity exploitation core is still effective during the EP stage, but it cannot be applied during the WG stage because it calculates matrix multiplication or convolution between IA and OE. Instead, pruned weight should be discarded and the calculation of the corresponding gradients can be skipped. It results in output sparsity during the WG stages. Output sparsity exploitation during the WG stage has big benefits thanks to both useless computation avoidance and memory access removal. For this reason, recent training processors [26, 31, 47, 78] supported triple sparsity exploitation for energy-efficient training with iterative pruning.

8.2.4 In- and Out-Slice Skipping

Dual or triple sparsity exploitation is proposed recently for zero-skipping-based DNN training, but they induce complex data-path leading to low efficiency at the dense data computing. In addition, the activation functions used for recent DNN architecture have been varied such as leaky-ReLU [52] and mish [53]. It wastes lots of energy if there is no sparsity during the DNN training. The work [29] focused on this drawback and utilized bit-slice (4-bit)-level sparsity induced by the Gaussian-like distribution. Moreover, it skipped partial accumulation-slices which should be truncated before it is used for the next layer. Thanks to the in- and out-slice skipping, it simplifies the sparsity exploitation data-path and utilizes slice-level sparsity in the entire training stages.

8.3 Bit-Precision Optimization

Even if the training processor adopts sparsity exploitation, its energy-efficiency is still lower than the conventional mobile inference processors because it requires high-bit-precision representation. Previously, it was considered that the DNN training requires 32-bit, but the required bit-precision is now continuously decreasing thanks to the researches on the low-bit-precision training. We will introduce recent researches to reduce the required precision during the training.

8.3.1 New Number Representation

The first trial to reduce bit-precision was the floating-point (FP) 16-bit IEEE 754 standard format. However, naïve FP16 based training degrades training accuracy because the exponent bit-width is too low to fully represent the weight-gradient. There were many substitutions proposed for better DNN training (Fig. 8.4).

New Floating-Point Representation

As shown in Fig. 8.5, there are three new FP representations that have wider data representation range thanks to the high-bit-width exponent. The bit-width of the exponent in Bfloat16 [14] and TensorFloat [54] is the same as the original FP32, so they show high compatibility with FP32. However, their energy efficiencies are still

Paper List	Sparsity Exploitation									Additional Sparsity -related Feature
	FF			EP			WG			
	IA	FW	OA	OE	BW	IE	IA	OE	G	
[48]	X	X	X	X	X	O	X	X	X	Stochastic error pruning
[30]	X	O	X	X	O	X	X	X	X	Group-sparse pruning
[47]	X	O	X	X	O	X	X	X	O	Early-training-step pruning
[50]	X	O	X	X	O	X	O	X	X	Continual weight decay for pruning
[21]	O	X	X	X	X	X	O	X	X	Feature Dividing for Mixed Precision
[24, 27]	O	X	O	X	X	O	O	X	X	Output sparsity prediction with MSBs (FXP) or exponent (FP)
[29]	O	X	O	O	X	O	O	X	X	In-out Slice Skipping for ReLU-robustness
[49]	O	O	X	X	O	X	O	X	X	-
[26, 31, 51, 78]	O	O	X	X	O	O	O	X	O	Adaptive in/out/weight skip + Structured Pruning

Fig. 8.4 Summary of sparsity exploitation DNN training processors

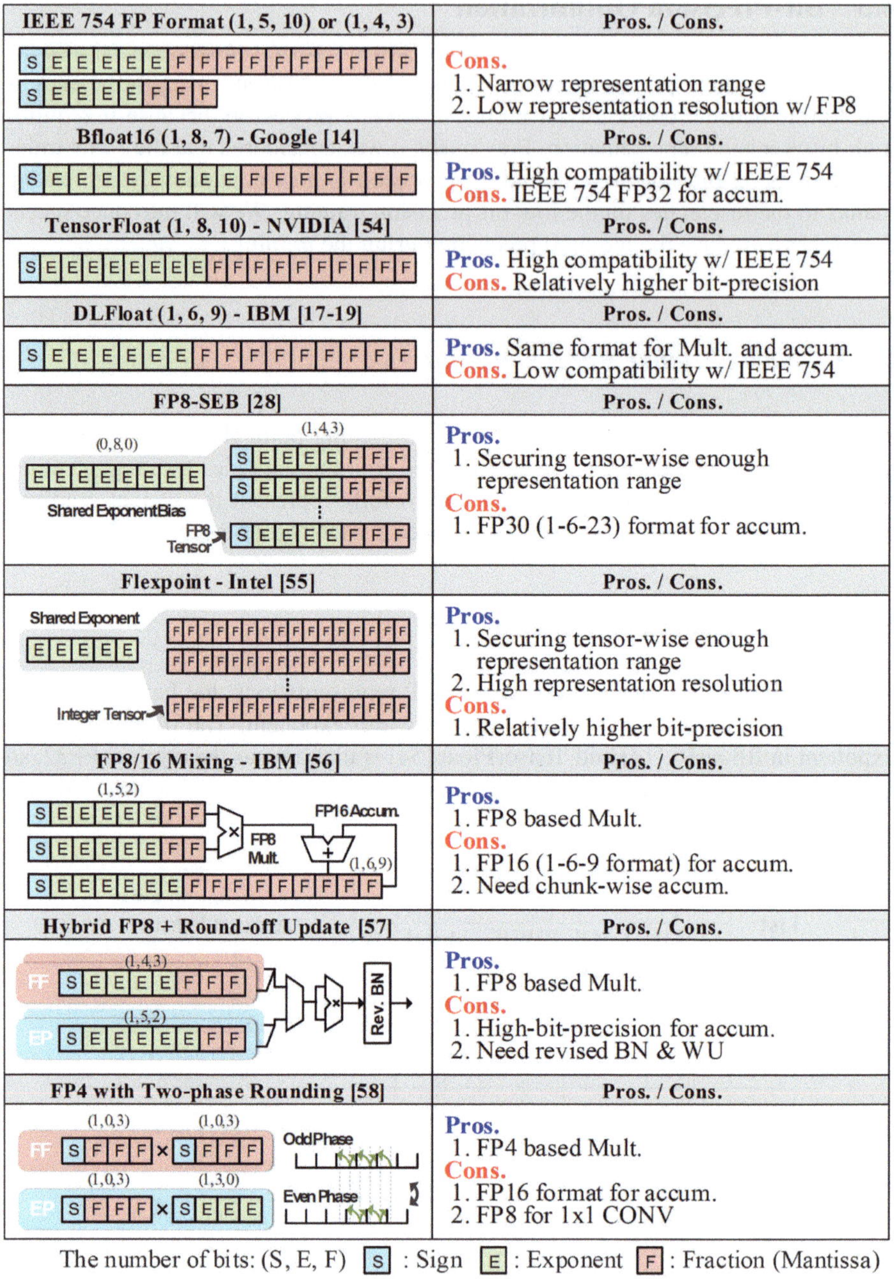

Fig. 8.5 Summary of floating-point (FP) representation for DNN training

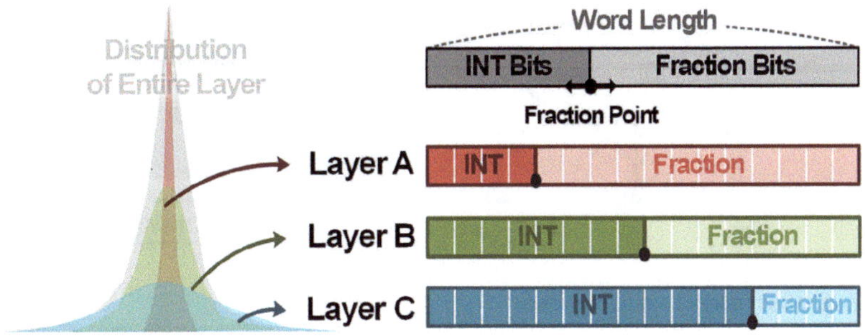

Fig. 8.6 Outline of dynamic fixed-point (DFXP) representation

low because they need more than 16-bit for data accumulation. The DLFloat [17–19] was proposed to unify the data representation method of both input operand and accumulation. Flexpoint [55] tried to substitute FP with FXP representation using a shared exponent management algorithm together for simplification of MAC design, but it failed to reduce the required bit-precision to less than 16-bit.

Nowadays, the FP based training can be realized with less than 16-bit by utilizing the narrow distribution that appears in each of the tensors [28], chunks [56], or training stages [57, 58]. The required representation range can be reduced by dividing computation through the shared exponent [28], chunk-wise accumulation [56], and hybrid representation among different training stages [57, 58]. Although the successful bit-precision reduction is observed during the multiplication, they still need high-bit-precision for accumulation and require additional hardware units for hybrid representation [57] or two-phase rounding [58] (Fig. 8.6).

New Fixed-Point Representation

Fixed-point (FXP) representation needed much higher bit-precision compared with the FP. The required bit-precision can be reduced dramatically when dynamic FXP (DFXP) [3, 29, 32] is adopted. It adjusts the required integer length to fit the layer-wise narrow distribution instead of considering entire layers. Even though the DFXP shows remarkable performance improvement, it still needs higher bit-precision (>8-bit) compared with the FP because the low-precision FXP has limited representation resolution. As shown in Fig. 8.7, stochastic rounding (SR [59]) was proposed to increase its virtual representation resolution. In addition to SR, stochastic thresholding (ST [29]) of Fig. 8.8 was co-designed to further increase representation resolution by loosening its overflow judgment and removing outlier data. Combination of SR and ST successfully demonstrated 43% bit-precision reduction in the CIFAR-100 dataset [29].

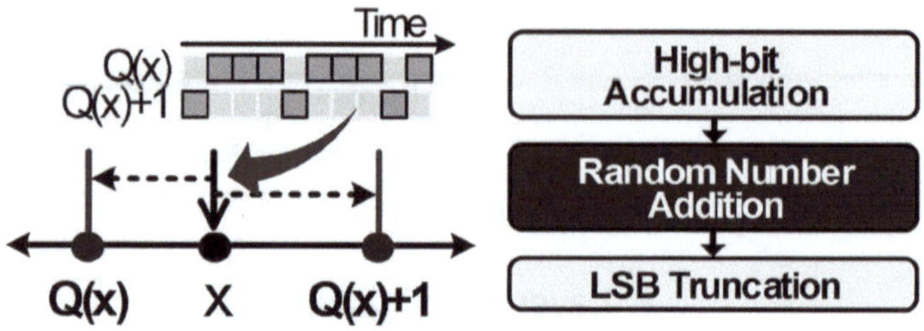

Fig. 8.7 Outline of stochastic rounding

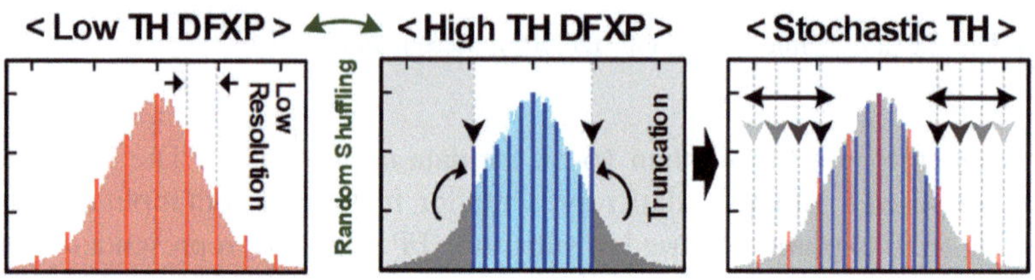

Fig. 8.8 Stochastic thresholding

8.3.2 Low-Bit-Precision Training Algorithm

Not only new number representation but also new algorithms were also proposed for low-bit-precision training. One of them is fine-grained mixed-precision (FGMP [21]), which divides IA into two different bit-width formats. The majority of data are placed near zero, so it is easy to be covered by the low-bit-precision data. However, ignoring precise representation of the outlier data degrades its training performance significantly. To solve this problem, the FGMP uses high bit-precision to represent only outlier data while it maintains low bit-precision for the majority of data.

Layer-wise adaptive precision scaling (LAPS [29]) further reduces the required bit-precision by using the automatic bit-precision tuning during the training. Required bit-precision can be varied if the training scenario is changed, but it is hard to predict the optimal bit-precision at the beginning of training. LAPS continuously monitors the similarity of high-precision and low-precision convolution results and increases the bit-width if the difference becomes larger. Bit-precision found by LAPS finally can be varied according to the dataset and network complexity.

8.3.3 Hardware Architecture for Bit-Precision Optimization

Hardware architecture and PE circuit should be modified to support the new number representations and low-bit-precision optimization algorithms.

Fused Multiply–Add Unit

To accommodate the new FP representations, the design of the basic MAC unit should be modified. Since the required bit-precision of the training is usually higher than the inference, training processors showed lower efficiency compared with the inference processors. Fused multiply–add (FMA) units can support both high-precision and low-precision computing and it becomes a common way to reduce the efficiency gap between inference and training processors.

Active Training Supporting Unit

Both FGMP [21] and DFXP [3, 29, 34, 35] need streaming data analysis units that can calculate the mean/variance or overflow ratio of the OA. Based on the analyzed tensor-wise statistics, the FGMP converts more than 90% of FP16 accumulation results to FP8 operands during the ResNet-18 training. In DFXP based computing, it increases the integer length of the corresponding layer if the overflow ratio exceeds the pre-determined threshold [3]. The LAPS [29] performed it in a peripheral unit that calculates the difference and counts only large differences. The new bit-precision is determined by the internal finite-state machine, and then it is applied from the next training iteration. The bit-precision reduction algorithms usually require an additional unit to monitor output distributions and give useful information for run-time bit-precision optimization.

8.4 Weight-Gradient-Update Stage Acceleration

As mentioned in Sect. 2.6.4, the WG stage shows memory-intensive computing which is a distinct characteristic compared with the other training stages. It results in two different design approaches of the DNN training core as described in Fig. 8.9.

Sparsity-aware WG acceleration assumes that it adopts weight pruning. Acceleration of the WG stage without weight pruning is still challenging because the bit-precision of the gradient and primal weight should be higher than the bit-precision used during the FF and EP stages. There is no practical realization of WG accelerator in current DNN training processors because aggressive approximation for the WG stage has the risk of losing training accuracy. However, optimization of the WG stage is essential for high performance training processor, so we summarize the existing architecture- or algorithm-level acceleration method of the WG stage.

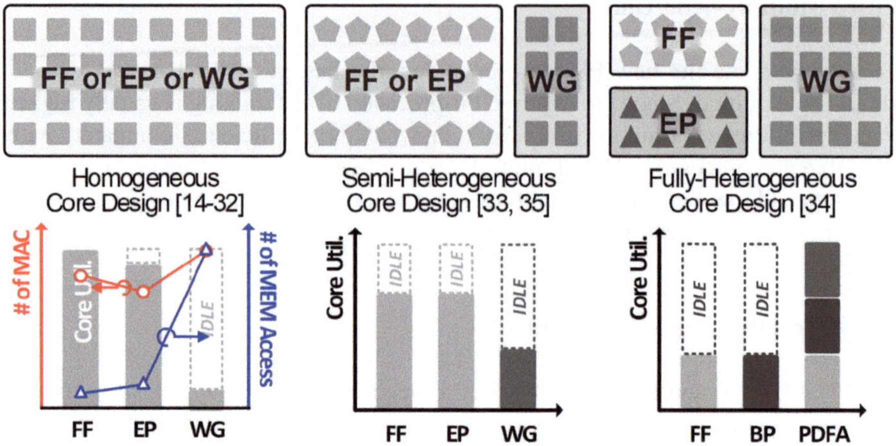

Fig. 8.9 Homogeneous and heterogeneous core design for DNN training

8.4.1 Homogeneous Core Approach

The majority of training processors [14–32] used the same core design to accelerate the three training stages. The homogeneous core shows low computing efficiency because it needs additional peripheral blocks and a complex data-path to meet the different requirements of the three different training stages. In spite of this disadvantage, core utilization can be dropped significantly if the role of the same cores is varied according to the three training stages. That is because the BP based training suffers from the backward locking problem as indicated in Sect. 2.6. Therefore, the homogeneous core design focuses on the processing time reduction of each layer by maximizing core utilization.

8.4.2 Heterogeneous Core Approach

Even though homogeneous core focused on maximizing core utilization, the authors in [33–35] adopted heterogeneous core for efficiency improvement. A heterogeneous core minimizes the required logic to support three different training stages, and a dedicated core accelerates each stage, especially for the WG stage to enhance computing efficiency. Although the separate WG core is useless during FF and EP stage, it can simplify FF and EP core architecture to maximize its computing efficiency during training. However, the heterogeneous core design shows low throughput because backward locking disturbs parallel processing. One processor [34] solved this problem by using its own backward unlocking method. It further improved computing efficiency by adopting a fully heterogeneous core design. Furthermore, it also realized high-speed training through the parallel processing of three training stages. The details of this solution will be discussed in Sects. 8.5 and 8.6.

8.4.3 Gradient Bit-Precision Reduction

Data compression is the most used WG acceleration method because the WG stage suffers from the memory-intensive computation. The authors in [60–62] suggested ultra-low-bit gradient quantization, but their main purpose was to reduce the gradient communication among the multiple edge devices during the distributed learning. Since the primal weight still needs full-precision (16-bit), the quantized gradient should be converted to the original bit-width during the weight update operation. In addition, a large portion of memory access is still required due to the data movement of the primal weight and momentum from the memory.

8.4.4 Sparse Gradient Selection

Gradient compression can be realized by not only quantization but also gradient sparsification. The work [63, 64] accumulated weight gradients every iteration and updated only when the accumulated gradient became larger than the predetermined threshold. These methodologies require a large internal buffer to store all weight gradients, which is not practical in training on micro-AI systems. Another sparsification [65] updated the weight only when it had a large gradient or the target confidence was low [66]. Large-gradient-only or low-confidence-only weight update showed great compression ratio, but they can cause a slow training curve. Furthermore, they are not easy to apply to other complicated DNN applications such as object detection and image-to-image generation.

8.4.5 Selective DNN Training

Although bit-precision reduction or sparsification during the WG stage shows higher efficiency than the naïve gradient calculation. It still requires a large buffer to store all intermediate IA and OE that appear during the FF and EP stages. The authors in [83] suggested bias-only fine-tuning which freezes weights and updates new bias to remove storage of intermediate IAs. Another method is selective layer training [32] which focused on a few layers but skips memorizing the IA and OE of the useless layers. Both methods can reduce the required memory footprint significantly but drop training quality due to aggressive gradient calculation skipping. The authors in [84] newly suggested channel-wise gradient skipping and also utilized both bias-only tuning and selective layer tuning to further optimize the training. To minimize the accuracy degradation, it predicts the accuracy loss and the reduced hardware cost during the compile time. Finally, selective training is a key enabler of on-device training by minimizing required memory access due to IA and OE.

8.5 Backward Unlocking

Consideration of backward locking is still challenging and there are few proposals to solve this problem. The authors in [68] proposed an out-of-order scheduling mechanism that performs inference of CL before the training of FCL is finished. However, balancing the workload between CL and FCL is challenging and FCL still suffers from backward locking problems. Backward unlocking (BU) without algorithm modification is difficult because BP itself has the serial computation characteristics.

8.5.1 Backward Unlocking with Delayed Back-Propagation

There are several backward unlocking methods [69, 70] using delayed BP. Decoupled delayed gradient (DDG, [69]) divides DNN into multiple groups and performs BP only within a group to loosen the backward locking problems. The gradient calculated by each group is transferred to the prior group at the next iteration. Since it adopts an inter-group pipeline during the BP, multiple gradients can be computed at the same time, but each group utilizes the FF results obtained at a different time step. Thus, it should copy the weights of the multiple different time-steps for parallel gradient computation.

Another method, fully decoupled gradient (FDG [70]), copies intermediate IAs instead of weights. Not only EP but also the FF stage is divided with the multiple groups and makes a big pipeline structure among them. Each group updates its weight with the delayed IAs and performs FF operation as soon as it completes the weight update. It can further decrease processing time thanks to the pipeline structure of both FF and EP stages. In spite of their success, both methods are considered only in the multi-GPU processing because they need a large buffer to store all weights or IAs of the delayed batches. Therefore, a memory-efficient backward unlocking method is necessary in the micro-AI systems.

8.5.2 Direct Feedback Alignment Based Approaches

Direct feedback alignment (DFA [71–74]) can propagate the errors of the last layer to the entire prior layers simultaneously. The problem of backward locking can be released in DFA-based training because the next inference can be started right after

	Related Figure	MEM of A&E	MEM of W	Proc. Time
BP [40]	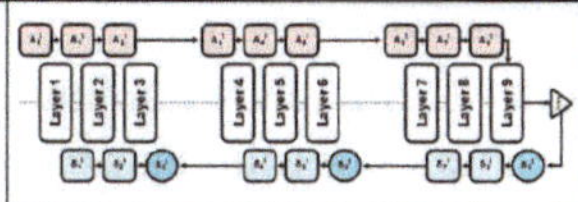	$A_0+A_1+A_2$ $+E_0+E_1+E_2$ $\approx \mathbf{A+E}$	$W_0+W_1+W_2$ $\approx \mathbf{W}$	$T_{A0}+T_{A1}+T_{A2}$ $+T_{E0}+T_{E1}+T_{E2}$ $\approx \mathbf{T_A+T_E}$
DDG [69]	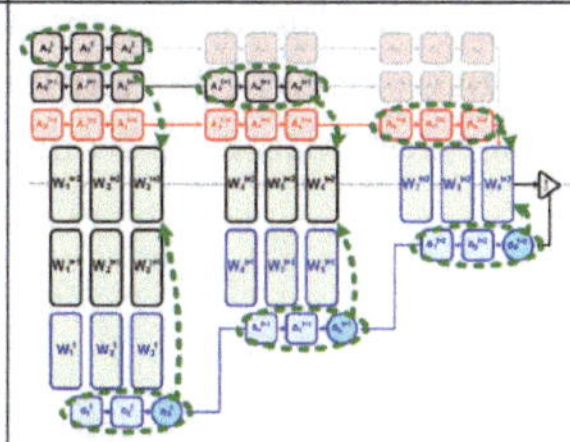	$3A_0+2A_1+A_2$ $+E_0+E_1+E_2$ $\approx \mathbf{2A+E}$	$3W_0+2W_1+1W_2$ $\approx \mathbf{2W}$ (W Access ▲)	$T_{A0}+T_{A1}+T_{A2}$ $+\max(T_{E0\sim E2})$ $\approx \mathbf{T_A+T_E/3}$
FDG [70]	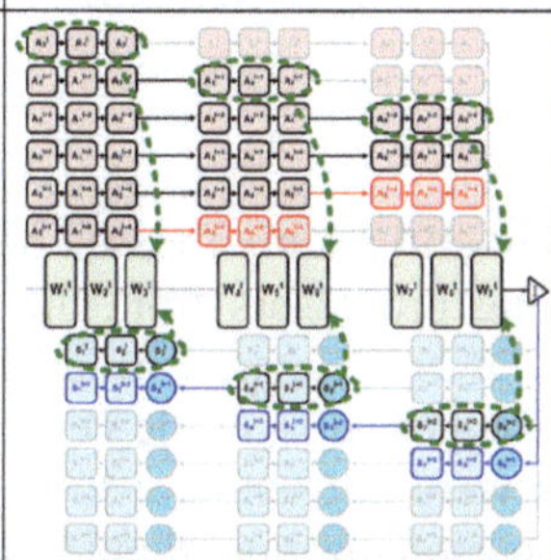	$6A_0+4A_1+2A_2$ $+E_0+E_1+E_2$ $\approx \mathbf{4A+E}$ (A Access ▲)	$W_0+W_1+W_2$ $\approx \mathbf{W}$	$\max(T_{A0\text{-}A2})$ $+\max(T_{E0\text{-}E2})$ $\approx \mathbf{(T_A+T_E)/3}$
DFA [71-74]	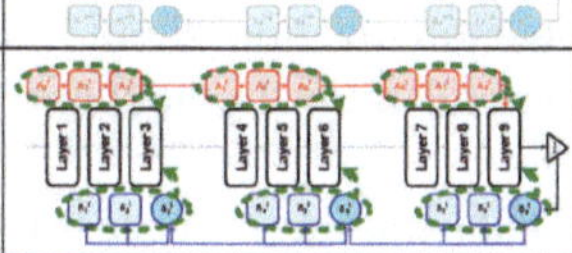	$A_0+A_1+A_2+E_{last}$ $\approx \mathbf{A+E_{last}}$	$W_0+W_1+W_2$ $+BW$ $\approx \mathbf{W+BW}$	$\max(T_{A0}, T_{E0})$ $+\max(T_{A1}, T_{E1})$ $+\max(T_{A2}, T_{E2})$ $\approx \mathbf{\max(T_A, T_E)}$

Fig. 8.10 Summary of typical backward unlocking methodologies

this direct error propagation. As shown in Fig. 8.10, it shows the minimum memory accesses compared with the other two backward unlocking methods. The only extra cost is that it needs an additional backward weight (BW) which is independent of the forward weight (FW). It now turns out that the overhead of the BW can be reduced by a maximum of 96.9% when using binarization [72] or sparsification [73]. Thanks to the lowest memory requirement, the DFA algorithm was selected by the recent on-chip training ASIC [34, 80], and it successfully demonstrated DFA-based online training in the object tracking or object detection application. It constructs a pipeline structure among the three training stages and also hide the latency caused by EP and WG. Thanks to the pipelined DFA (PDFA), it showed at least 2× higher training speed compared with the BP based acceleration.

8.6 DNN Training Accelerator Examples

In this section, the real chip implementation results of the DNN training processor will be discussed. Firstly, we will introduce the examples of training processors and clarify their target training applications. The reason why the algorithm co-design is necessary for mobile-oriented DNN training processor will be explained with practical examples. Lastly, key differences in accelerator design between inference and training will be discussed.

8.6.1 Applications and Examples of Training Processor

Applications of DNN Training

Training processors reported from the industry [14–19] mainly focused on general-purpose DNN training. On the contrary, training processors from the academy [20–35] targeted local training which fine-tunes DNN to be more accurate in user-specific datasets. In situ personalization is also suggested to fit the network to the user-specific tasks. However, these applications were not sensitive to the training speed because local dataset optimization and personalization should collect the user's behaviors and experiences for a long time and train the network when the device is not in use, usually in midnight.

Super-fast training is required when online DNN training is necessary. One of the ASICs [34] showed an example that utilized online training to extend the role of object classification networks to object tracking. It learned the shape of the first bounding box and decided whether the surrounding boxes were similar to the target. Since it updates the network to learn the continuously moving target, it can achieve robust tracking under the shape deformation and illumination changes. Another example was TKD suggested by [29, 42]. It realized energy-efficient but accurate object detection by adopting online knowledge distillation to the lightweight DNN. The lightweight network itself has been considered impractical due to its poor performance. This problem can be solved if the network is trained by data with labels generated by heavyweight teacher network inference. It needs the inference of a heavy teacher network, but this process is performed infrequently and consumes significantly less energy compared to the direct use of the teacher network. These examples show that DNN training can be used as not only long-term user customization but also short-term domain adaptation to further improve its functionality and performance.

Examples of Training Processor Design

The majority of processors [14–32] adopted homogeneous core design, and they were programmable to be applied in various types of networks and applications.

Moreover, they also emphasized a new FP-number-representation-based PE design by utilizing a new number representation method and adopted FMA for energy-efficient inference and training [15–19, 21, 22, 26, 138]. Mobile training processors are mainly developed by the academy and they were continuously evolved. The first-generation mobile training processors [20, 22, 23, 25, 35] focused on training itself by supporting transpose-read and reconfigurable data-path. The second-generation processors [21, 27, 31, 138] were designed by combining sparsity exploitation or low-bit-precision training schemes to improve its throughput and efficiency. It is now evolved into third-generation processors [26, 29, 30] which further improved the performance by proposing the adaptive methods of bit-precision or pruning ratio control even when the training is in progress.

8.6.2 Algorithm–Hardware Co-optimization in Training Processor

Most of the training processors shown in Fig. 8.11 adopted algorithm–hardware co-design, and it now becomes an inevitable trend for realizing DNN training at the micro-AI systems. Since the naïve BP based training requires large hardware resources, the algorithmic co-optimization is essential for high-speed and energy-efficient training.

New Algorithm and Its Dedicated Hardware

Some algorithms are not that efficient in conventional DNN accelerators and require completely different hardware designs as shown in Figs. 8.12, 8.13, and 8.14. Nevertheless, when they are co-designed with their optimized hardware, it shows much higher efficiency compared with the naïve implementation of DNN training.

The GANPU [138] of Fig. 8.12 sought co-optimization by proposing a hardware-friendly output sparsity prediction algorithm for the FP number system to support fast training. Sign and exponent bits, which represent the scale of data, are used to speculate whether the ReLU result will be zero. Given that multiplications between exponents are mathematically identical to integer additions, a dedicated speculation unit was able to be integrated on-chip with minimal resource overhead, and it helped the main convolution core to exploit both input and output sparsity simultaneously.

The PNPU [26] of Fig. 8.13 proposed a new iterative pruning algorithm that utilized both coarse-level and fine-level pruning for efficient training acceleration. It approximated similarity calculation among the output channels through random channel sampling. The PNPU designed its dedicated pruning core and constructed a pipeline structure between the main core and the pruning core to support iterative pruning without PE utilization drop. In addition to the pruning unit, the main core

Papers	Process	Verified Task	Application	PE Unit	Bit Reconfig.	Core Arch.	Sparsity Exploitation	Prec. Optimization	Etc.
Google TPU-v2 [14]	20nm	GP	-	bfloat16 FPU	-	Homogeneous	-	bfloat Representation	-
GTX 1080Ti [15]	16nm	GP	-	FP16/32 FMA + FP64 FPU	16,32,64-bit	Homogeneous	-	FP16-FP32 Mixed-prec. Training	-
NVIDIA V100 [16]	12nm	GP	-	FP-FXP FMA	FP: 16,32,64-bit FXP: 4,8-bit	Homogeneous	-	FP16-FP32 Mixed-prec. Training	-
S.VLSI'18 [17]	14nm	GP	-	FP16 FPU	-	Homogeneous (+SFU)	-	Supporting Binary/Ternary INF	-
S.VLSI'20 [18]	14nm	GP	-	DLfloat16-FP32 FMA	16,32-bit	Homogeneous (+SFU)	-	DLFloat16	-
ISSCC'21 [19]	7nm	GP	-	FP-FXP FMA	FP: 8,16,32-bit FXP: 2,4-bit	Homogeneous (+SFU)	-	-	Workload-aware Throttling
S.VLSI'18 [33]	65nm	Classification	Local Training	5b LUT-Idx 8b×8b MAC	-	Heterogeneous (CL⇔FCL)	IA/W Sparsity (ReLU, Pruning)	-	FCL Training by Quant. Center Tuning
ISSCC'19 [20]	65nm	DRL	DRL	bfloat16×FXP4 MAC	W: 4,8,16-bit	Homogeneous	-	-	RL Experience Compression
ISSCC'19 [21]	65nm	Classification	Local Training	FP8-16 FMA	8,16-bit	Homogeneous	IA Sparsity (ReLU)	FGMP	Input Load Balancer
S.VLSI'19 [34]	65nm	Tracking	Tracking	CL: FP8 FCL: FXP13/16	-	Heterogeneous (CL⇔FF⇔EP⇔WG)	CL:IA Sparsity (ReLU)	Binary Backward Weight	Pipelined DFA
ASSCC'19 [22]	40nm	-	Local Training	FP-FXP FMA	FP: 16-bit FXP: 5,10-bit	Homogeneous	-	-	Channel Re-arrangement
SSCL'19 [23]	65nm	Classification	Local Training	FXP16 MAC	-	Homogeneous	-	-	Diagonal Storage Pattern
ISSCC'20 [24]	65nm	GAN	Personalization	FP8-16 FMA	8,16-bit	Homogeneous (+EORS unit)	IA/OA Sparsity (ReLU)	-	Exponent-only ReLU Speculation (EORS)
ISSCC'20 [25]	28nm	-	Local Training	(FXP) 2b×1b Multiply cell	2,4,8-bit	-	-	-	Custom Transpose SRAM Mixed-mode Computing
S.VLSI'20 [26]	65nm	Classification	Local Training	FP8-16 FMA	8,16-bit	Homogeneous	IA/OA/W Sparsity (ReLU, Pruning)	-	(Iterative) Stochastic Coarse-Fine Level Pruning
JSSC'20 [35]	65nm	Classification	Personalization	DFXP8 MAC	-	Heterogeneous (FF/EP⇔WG)	-	DFXP	Masking Logic for ReLU/Pooling
JSSC'20 [27]	28nm	Classification	Local Training	FXP2 MAC	2,4,8-bit	Homogeneous	IA/OA Sparsity (ReLU)	-	Bidirectional Speculation + QVF Opt. w/ Q-learning
ISSCC'21 [28]	40nm	Classification	Local Training	FP8 FPU + SEB	-	Homogeneous	-	FP8 + Shared Exponent Bias	-
JSSC'21 [29]	28nm	Classification + Detection	Local Training +TKD [42]	DFXP5 MAC	4,8,12,16-bit	Homogeneous	In-/Out-slice Sparsity (Near-zero., ACC)	SDFXP + LAPS	-
S.VLSI'21 [30]	28nm	DRL	DRL	bfloat16 MAC	16-bit	Homogeneous	W Sparsity (Pruning)	-	Group-sparse Training + Sparse W Transposer
S.VLSI'21 [31]	28nm	Classification	Local Training	FP-FXP FMA	FP: 8,16-bit FXP: 4,8-bit	Homogeneous	IA/OA/W Sparsity (ReLU, Pruning)	-	Redundancy Speculation + BN Reformulation
JSSC'21 [32]	65nm	GAN	Personalization	FP16 MAC	16-bit	Homogeneous	-	-	Selective Layer Re-Training + Reordering Instance Norm.

[ ]: Industry [ ]: Academy [Green Letters]: Algorithm Co-Optimization

Fig. 8.11 Summary of DNN training processors

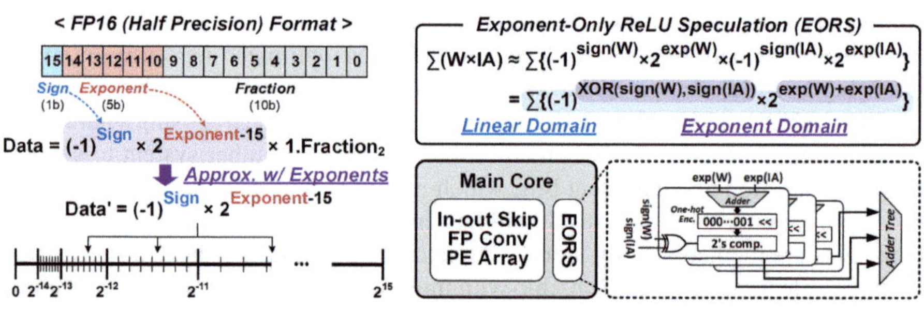

Fig. 8.12 GANPU [138]: exponent-only ReLU speculation algorithm and adder-only light convolution supporting hardware

is designed to support triple sparsity exploitation and its hardware complexity could be reduced by utilizing coarse-zero skipping.

Pruning-aware training processors [26, 31] were efficient only when the pruning ratio became high enough so that their performance could be degraded significantly

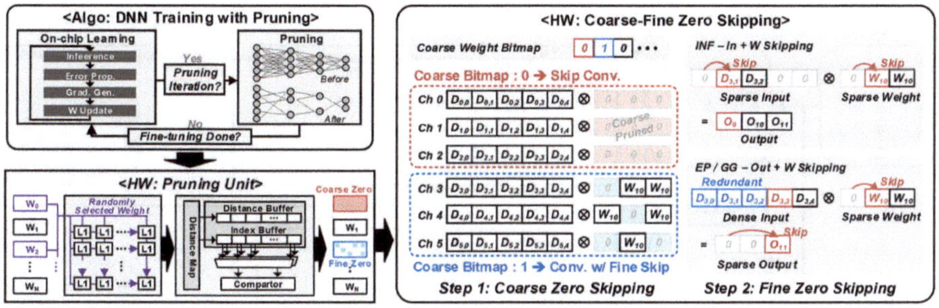

Fig. 8.13 PNPU [26]: new iterative pruning algorithm and core architecture to exploit both coarse and fine-grained sparsity

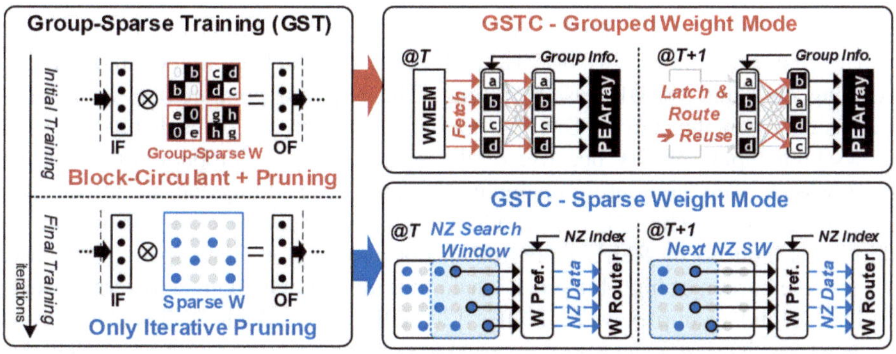

Fig. 8.14 OmniDRL [30]: group-sparse training algorithm with reconfigurable core architecture to support both weight reuse and skipping

at the beginning of training. It is because they could not remove the weight connection before the saliency of weight was determined. OmniDRL [30] of Fig. 8.14 solved this problem by utilizing block-circulant-based weight grouping and the pruning of small weights together in the initial training phase. The new algorithm, group-sparse training (GST), induced both repeated patterns and sparsity in the weight. Group-sparse training core (GSTC) suggested by [30] improved training performance by enabling both grouped weight reuse and zero-weight skipping through a weight router and pre-fetcher.

Hardware Performance Improvement Through the Algorithm Combination

Algorithms introduced in Figs. 8.12 and 8.14 were newly proposed for energy-efficient DNN training acceleration, but it is effective only when its dedicated hardware exists. Unlike these examples, there were ASICs that have greater synergy when the algorithm is combined.

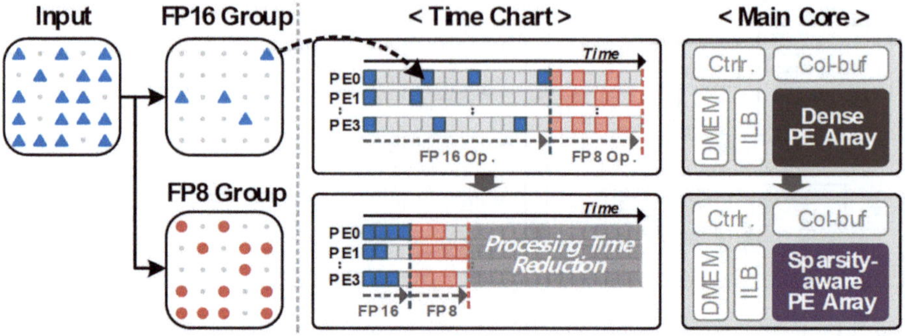

Fig. 8.15 LNPU [21]: Combining IA zero-skipping core and FGMP

The first example was LNPU [21] of Fig. 8.15 which combined input zero-skipping core and FGMP based training. The required computation is linearly increasing when the naïve FGMP is applied. However, IA of each bit-precision shows high input sparsity because of the zero paddings. Computation including zeros wastes many computing resources because zero paddings do not affect final computing results. The LNPU [21] adopted the input sparsity exploitation core to skip zeros induced by ReLU or FGMP. It not only exploits data sparsity but also reduces the burden of high-bit-precision computing to obtain a great synergy effect.

The second examples were ASICs [27, 29] which adopted bit-scalable core architecture. Bit-precision can be varied according to layer number so that the bit-scalable core architecture could remove useless operations during low-bit-precision training and maximize both training speed and efficiency. In addition to these purposes, it was able to realize an ReLU prediction algorithm [27] by performing MSB-only computing. After MSB-only prediction, it skipped the LSB computations of which results would have the zero-expected value. Output slice skipping [29] was another algorithm that shows synergy with the bit-scalable architecture. It could compensate for the throughput degradation problem during high-precision computing by excluding LSB accumulations. Lastly, it could be combined with the run-time bit-precision optimization algorithm, LAPS, as shown in Fig. 8.16. The HNPU [29] minimized the overhead of difference calculation between high-precision and low-precision convolution results by performing only LSB accumulations. The bit-scalable architecture used in [29] finally minimized the extra time required for bit-precision searching and optimized training performance with automatically discovered precision number.

The last example is DF-LNPU [34] of Fig. 8.17 which adopted a heterogeneous core design. The heterogeneous core design can optimize each training stage and minimize hardware resources to support training but has the problem of core utilization drop. The DF-LNPU [34] adopted a heterogeneous core design but no longer suffered from low core utilization problems by utilizing PDFA based backward unlocking. The PDFA based optimization enabled parallel processing of three different training stages. Combination with backward unlocking algorithm

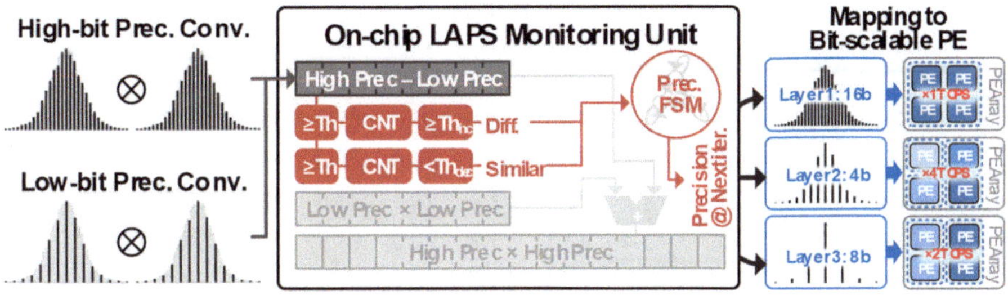

Fig. 8.16 HNPU [29]: combining bit-scalable core and LAPS

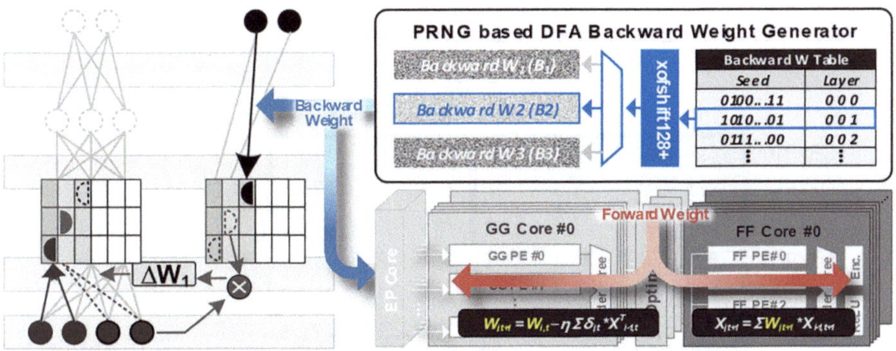

Fig. 8.17 DF-LNPU [34]: combining fully heterogeneous core and PDFA

compensated for throughput degradation caused by backward locking while retaining the advantage of the heterogeneous core design.

The main challenges illustrated in Sect. 2.6 can be solved when the algorithm is co-designed with the training hardware. These examples clarify that algorithm–hardware co-optimization plays an important role in realizing DNN training at the edge device, even with limited hardware resources.

Relationship of Sparsity and Quantization During DNN Training

Sections 8.2 and 8.3 introduce typical examples of sparsity exploitation and quantization method for on-device DNN training. In this section, we will discuss the relationship among these optimization techniques. We construct the experiment of three different ResNet trainings in the CIFAR-100 dataset. In this experiment, we pre-trained the network with FP32 and retrained them after weight pruning and DFXP based quantization. During the quantization, we modified bit-precision of weight, IA, and OE but maintains high accumulation bit-width. To analyze the relationship between pruning and quantization, we introduce a new graphical display, PQ (pruning quantization) plot, which visualizes the test accuracy according to varying training conditions to help designers to select optimal pruning and

Pruning Ratio	0.0	0.1	0.2	0.3	0.4	0.5	0.6	0.7	0.8	0.9
DFXP16	74.0	73.8	72.9	72.1	71.1	69.7	67.7	63.6	43.6	27.5
DFXP14	74.0	73.4	72.5	71.8	70.5	69.6	67.7	63.3	53.0	33.4
DFXP12	73.3	73.1	72.7	71.9	71.5	70.0	68.0	63.0	43.5	33.0
DFXP10	73.3	73.4	72.7	72.3	71.5	70.1	67.6	63.3	51.0	29.3
DFXP8	73.9	73.7	73.4	72.5	71.7	70.0	67.5	63.4	54.1	39.2
DFXP6	73.1	72.7	72.3	71.9	70.6	69.7	67.1	61.9	52.3	30.3
DFXP4	32.3	31.3	28.0	23.6	24.7	21.5	13.3	14.2	5.4	2.1

Fig. 8.18 PQ plot of ResNet-9 (green box: >73.0% and blue box: >77.0%): test accuracy of ResNet-9 after re-training for pruning and quantization

Pruning Ratio	0.0	0.1	0.2	0.3	0.4	0.5	0.6	0.7	0.8	0.9
DFXP16	78.8	78.4	77.9	77.1	76.6	75.1	73.7	71.4	64.1	50.1
DFXP14	78.8	78.4	78.0	77.1	76.1	75.0	73.8	71.9	65.0	50.4
DFXP12	78.6	78.5	77.8	77.4	76.4	75.1	74.2	71.3	65.3	50.0
DFXP10	78.8	78.7	78.2	77.6	76.6	75.5	74.0	71.4	65.6	49.8
DFXP8	78.5	78.1	77.9	76.9	76.5	75.3	74.0	71.2	64.5	48.0
DFXP6	77.7	77.4	76.9	76.3	75.4	74.7	72.8	70.5	63.3	47.0
DFXP4	44.9	44.3	46.3	43.7	42.7	40.9	39.0	34.7	34.7	27.2

Fig. 8.19 PQ plot of ResNet-18 (green box: >73.0% and blue box: >77.0%): test accuracy of ResNet-18 after re-training for pruning and quantization

Pruning Ratio	0.0	0.1	0.2	0.3	0.4	0.5	0.6	0.7	0.8	0.9
DFXP16	78.5	78.4	78.0	77.9	77.4	76.6	75.8	73.6	70.0	54.4
DFXP14	78.4	78.3	78.3	77.7	77.5	76.8	75.5	74.2	69.5	54.5
DFXP12	78.6	78.6	78.4	77.3	77.3	76.6	75.4	73.7	69.2	54.1
DFXP10	78.1	78.1	78.3	77.5	77.4	76.5	75.6	73.7	70.6	55.3
DFXP8	78.5	78.5	78.2	78.0	77.5	76.6	75.8	73.6	68.6	52.1
DFXP6	78.3	77.9	77.7	77.5	76.8	75.5	74.5	73.0	68.7	51.5
DFXP4	62.4	63.3	61.8	62.7	62.8	59.9	57.4	54.0	40.6	15.9

Fig. 8.20 PQ plot of ResNet-34 (green box: >73.0% and blue box: >77.0%): test accuracy of ResNet-34 after re-training for pruning and quantization

quantization level during the on-device training. Figures 8.18, 8.19, 8.20 shows three PQ plots obtained from our experiment. Green and blue boxes in Figs. 8.18, 8.19, 8.20 indicate acceptable conditions with two different accuracy thresholds. Within the accuracy boundary, the on-device training designer can choose the best condition based on the hardware characteristics of the training processor.

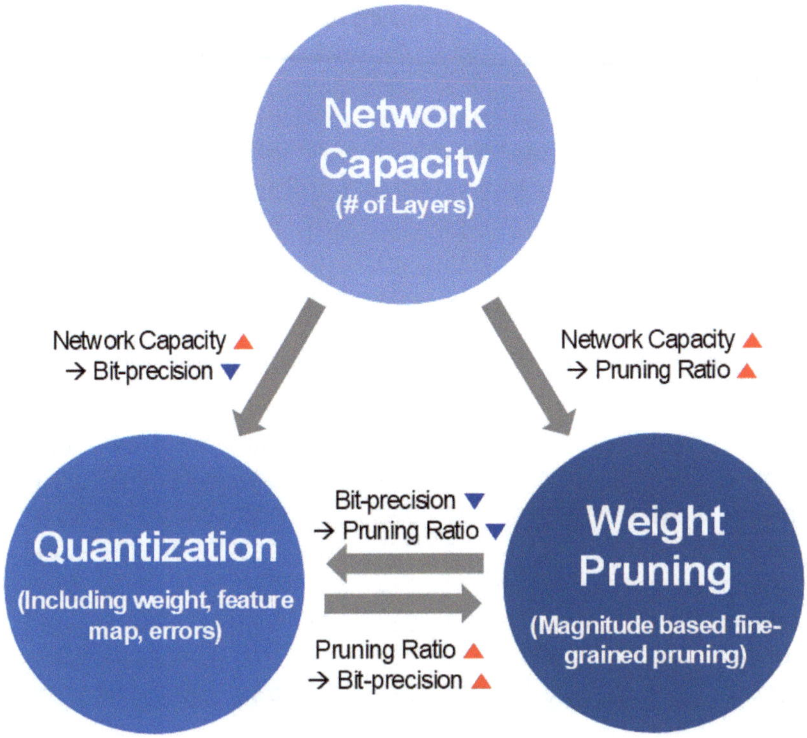

Fig. 8.21 Trade-off between weight pruning and quantization

There are two key observations obtained from the PQ plot as summarized in Fig. 8.21. First, DNN shows a low pruning ratio and high bit-precision if the network capacity is small. Second, there is a trade-off relationship between pruning and quantization. If a designer selects a high pruning ratio, high bit-precision must be used to maintain the original high accuracy. The optimal design point can be varied whether the target processor has the functionality of zero-weight skipping or bit-reconfigurability.

Another minor observation is about IA sparsity. Figure 8.22, 8.23, 8.24 summarizes the layer-wise average IA sparsity after pruning or quantization. As shown in this figure, IA sparsity increases when it has a higher pruning ratio or lower bit-precision. However, the IA sparsity within the accuracy boundary (indicated as a green or a blue box) shows a small variance, and thus, there is no expectation of additional throughput improvement through input sparsity exploitation when considering both pruning and quantization.

Figures 8.25, 8.26 shows an example of a DNN training execution comparison. We compare the ideal performance of the two DNN accelerators which support both weight skipping and bit-reconfigurability but have different bit-granularity. The optimal design point of the coarse-grained bit-reconfigurable accelerator shows a

	Pruning Ratio									
	0.0	0.1	0.2	0.3	0.4	0.5	0.6	0.7	0.8	0.9
DFXP16	0.53	0.53	0.56	0.58	0.61	0.65	0.69	0.72	0.78	0.82
DFXP14	0.52	0.53	0.55	0.58	0.62	0.65	0.68	0.72	0.77	0.82
DFXP12	0.52	0.53	0.55	0.58	0.61	0.65	0.69	0.73	0.78	0.82
DFXP10	0.52	0.53	0.55	0.59	0.62	0.65	0.69	0.73	0.79	0.82
DFXP8	0.53	0.54	0.56	0.59	0.62	0.66	0.69	0.73	0.78	0.82
DFXP6	0.57	0.61	0.63	0.65	0.67	0.7	0.72	0.76	0.79	0.82
DFXP4	0.76	0.78	0.76	0.79	0.79	0.79	0.8	0.83	0.84	0.85

Fig. 8.22 IA sparsity of ResNet-9 after re-training for pruning and quantization

	Pruning Ratio									
	0.0	0.1	0.2	0.3	0.4	0.5	0.6	0.7	0.8	0.9
DFXP16	0.59	0.6	0.6	0.62	0.63	0.66	0.69	0.73	0.79	0.84
DFXP14	0.59	0.6	0.61	0.62	0.63	0.66	0.68	0.73	0.78	0.84
DFXP12	0.58	0.6	0.6	0.62	0.63	0.66	0.69	0.73	0.78	0.84
DFXP10	0.59	0.6	0.61	0.62	0.64	0.66	0.69	0.73	0.78	0.84
DFXP8	0.61	0.61	0.62	0.63	0.64	0.66	0.69	0.73	0.79	0.85
DFXP6	0.65	0.65	0.66	0.67	0.67	0.69	0.73	0.76	0.81	0.85
DFXP4	0.83	0.83	0.83	0.84	0.85	0.86	0.86	0.87	0.89	0.91

Fig. 8.23 IA sparsity of ResNet-18 after re-training for pruning and quantization

	Pruning Ratio									
	0.0	0.1	0.2	0.3	0.4	0.5	0.6	0.7	0.8	0.9
DFXP16	0.6	0.61	0.63	0.64	0.66	0.68	0.7	0.74	0.79	0.84
DFXP14	0.6	0.61	0.63	0.64	0.66	0.68	0.7	0.73	0.78	0.84
DFXP12	0.6	0.61	0.62	0.64	0.65	0.67	0.7	0.73	0.78	0.84
DFXP10	0.61	0.62	0.63	0.65	0.67	0.68	0.71	0.74	0.78	0.84
DFXP8	0.62	0.63	0.64	0.66	0.67	0.69	0.71	0.75	0.79	0.84
DFXP6	0.68	0.69	0.7	0.71	0.72	0.73	0.75	0.78	0.81	0.86
DFXP4	0.85	0.85	0.85	0.86	0.87	0.87	0.89	0.9	0.91	0.94

Fig. 8.24 IA sparsity of ResNet-34 after re-training for pruning and quantization

higher pruning ratio than the fine-grained bit-reconfigurable accelerator (FGBRA). Instead, the FGBRA shows faster training in this experiment even with the lower pruning ratio. To sum up, the PQ plot can be useful for training processor designers to select optimal design hyperparameters. Utilizing both pruning and quantization is the best way to improve efficiency but needs to avoid lopsided optimization.

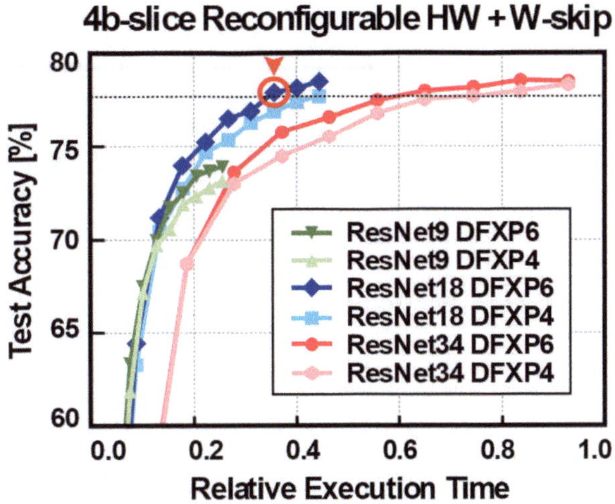

Fig. 8.25 ResNet re-training experiment in CIFAR-100 dataset w/DNN accelerator which has 4b-slice reconfigurability and zero-weight skipping

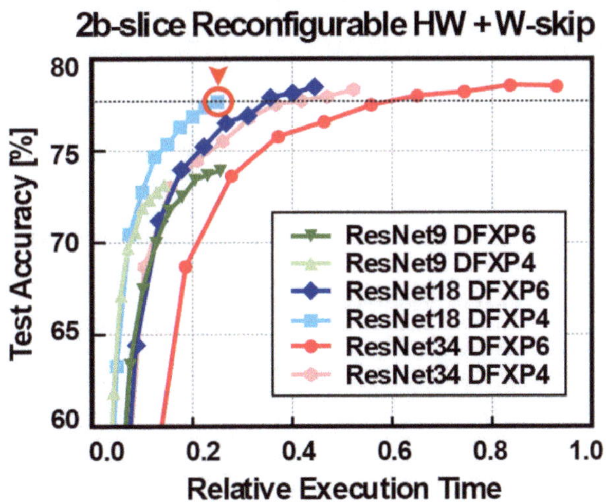

Fig. 8.26 ResNet re-training experiment in CIFAR-100 dataset w/DNN accelerator which has 2b-slice reconfigurability and zero-weight skipping

Design Comparison: HNPU-V1 and HNPU-V2

As illustrated in Sect. 1.2, the design of the on-device training processor can be varied according to their target training scenarios. There were two typical processors, HNPU-V1 [29] and HNPU-V2 [80], which utilized the same amount of hardware resources but were optimized differently due to their target application. Their key characteristics are summarized and the differences are highlighted in Fig. 8.27. The HNPU-V1 mainly focused on long-term and medium-term DNN training, and thus, it introduced automatic precision search during the training for the minimization of total training time and energy consumption. Moreover, it combined not only DFXP and SR but also ST to make a stable training curve even with the precision change caused by the LAPS. On the contrary, the HNPU-

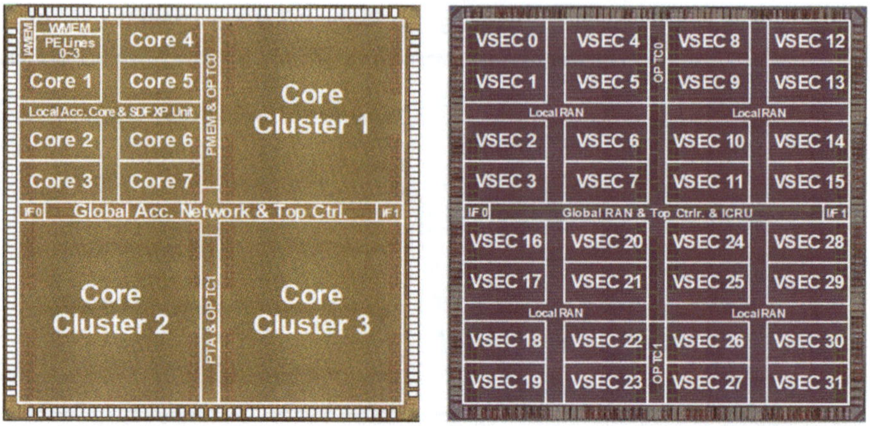

HNPU-V1 [29]	Processor	HNPU-V2 [80]
Training from the Scratch & Personalization	Target Application	Run-time Online DNN Tuning @ Object Detection
Samsung 28nm 1P8M CMOS	Technology	Samsung 28nm 1P8M CMOS
3.6mm × 3.6mm	Die Area	3.6mm × 3.6mm
0.58V (@ 2MHz) ~ 1.04V (@ 250MHz)	Op. Condition	0.63V (@ 10MHz) ~ 1.0V (@ 250MHz)
DFXP+SR+ST	Data Type	DFXP+SR
(4/8/12/16)-bit × (4/8/12/16)-bit	Data Bit-precision	(4/8/12/16)-bit × (4/8/12/16)-bit
X	Pruning-aware Training	O (Channel-pruning)
X	Backward Unlocking	O
FF: In-out Slice Skip EP: In-out Slice Skip WG: In-slice Skip	Sparsity Exploitation	FF: In-slice Skip + FW Skip EP: In-slice Skip + BW Skip WG: In-slice Skip + G Skip
LAPS	Advanced Feature	Low-cost SR
955-to1162 mW	Power	850-to-1032 mW
4.53 TOPS	Peak Performance	7.07 TOPS
4.74 TOPS/W	Energy Efficiency	7.89 TOPS/W

Fig. 8.27 Examples of algorithm–hardware co-design in DNN training ASICs: HNPU-V1 and HNPU-V2

V2 targeted short-term DNN training, which performs online tuning for real-world environmental adaptation. It adopted the pre-determined static bit-precision during the online DNN tuning and removed ST-related circuits in its PE architecture. The higher bit-precision requirement of HNPU-V2 due to input data uncertainty could disturb the realization of low-latency training. To resolve this problem, it adopted pruning-aware training and DFA based backward unlocking to reduce training latency. Since it does not require ST logic, it further optimized SR circuits by using the low-cost random number generator. In the same online tuning scenario, the HNPU-V2 can achieve 56.1% higher throughput and 66.5% higher energy-efficiency than HNPU-V1. Although the training method of HNPU-V2 can be lossy to be used in other training tasks, it successfully realized low-latency DNN tuning for fast environmental adaptation. In contrast, the HNPU-V1 can support any long-term or medium-term training without accuracy loss by adopting changing precision during the training. Thanks to precision search, it can minimize required internal/external memory access compared with HNPU-V2 while training most of the network without loss. As shown in this example, the design methodology of the DNN training processor can be varied according to the target application, which may require different optimization schemes.

8.7 Comparison of Inference and Training Processor Design

Figure 8.28 compares the main features of inference and training processors. Conventional inference processors focus only on the fast and efficient execution of well-trained networks while they do not consider whether the training of the network becomes slow or not. In contrast, training processor should achieve high performance in both inference and training without compromising the training accuracy.

<table>
<tr><th>Type</th><th>Inference Processor</th><th>Training Processor</th></tr>
<tr><td>Purpose</td><td>• Fast / Efficient Inference
• (Training can be slow)</td><td>• Training for Inference
• Fast / Efficient Training</td></tr>
<tr><td rowspan="2">Algorithm Level Feature</td><td colspan="2">• Fine-grained Mixed-precision [21]
• Sparsity Encoding: Zero-length Encoding [21] / CSC [33, 47, 48] / Bitmap [24]</td></tr>
<tr><td>• Binary/Ternary Quant. [9, 10]
• DNN Early-stopping [75, 76]
• Output ReLU Speculation [24, 27]</td><td>• Backward Unlocking [69-74]
• Quant-aware Training [27, 29]
• Pruning-aware Training [26, 31, 47]
• Selective Layer Training [32]</td></tr>
<tr><td rowspan="2">Architecture Level Feature</td><td colspan="2">• Input or Weight Zero Skipping [13, 21, 24, 26, 27, 29, 30, 31, 34, 47-51]
• Distributed Memory Architecture / Systolic Array [14, 32]</td></tr>
<tr><td>• Fused-layer based Acceleration [12, 75]</td><td>• Output Zero Skipping [24, 26, 27, 29, 31, 49]
• Training-stage-pipeline [34, 68]
• Transposable PE Array [20]
• Sparse Weight Transposer [30]
• Norm. Layer Reform [31, 32]</td></tr>
<tr><td rowspan="2">PE/Circuit Level Feature</td><td colspan="2">• Dynamic Fixed-point MAC [3, 29, 34, 35, 43]
• Voltage-Frequency Scaling</td></tr>
<tr><td>• Look-up Table based PE [3, 4, 10]
• Mixed-mode Computing [5, 8, 9]</td><td>• Stochastic Rounding Circuit [29]
• Binary BW Computing MAC [34]
• FP-FXP Fused Multiply-Add Unit [16, 18, 19, 21, 22, 24, 26, 31]</td></tr>
</table>

Fig. 8.28 Comparison of DNN inference and training processor

8.7.1 PE/Circuit-Level Differences

The inference processor can take advantage of ultra-low-bit quantization because the network can be re-trained multiple times in advance to compensate for accuracy degradation. Ultra-low-bit quantization enables look-up table based computing [3, 4, 10] for energy-efficient inference acceleration. Furthermore, an inference processor can utilize analog-domain computing [8] if it repeats re-training after adding noise components to a pre-well-trained network. However, both ultra-low-bit quantization and mixed-mode computing can cause accuracy degradation during the DNN training and slow down the training curve. For this reason, the majority of training processors adopt high-bit-precision FP based FMA instead. It sometimes uses backward weight binarization [34, 43] but only in limited applications.

8.7.2 Architecture-Level Differences

The training processor shows differences also in the architecture-level features. In the inference processor, they sometimes adopt fused layer-based acceleration [12, 75] to remove external memory access that appears during layer-by-layer computing. The fused layer is less effective in the training processor because all intermediate IAs and OEs should be stored and re-loaded at the WG stage. Moreover, the training processor shows more complex data-path architecture due to the transpose-read of weights or WG operation. It utilizes additional features such as transposable PE array [20] and output sparsity exploitation [26, 27, 31, 138] to compensate for efficiency drop caused by complex data-path.

8.7.3 Algorithm-Level Differences

Training processors also show different design methods from inference processors during the algorithm–hardware co-optimization. Recent application-specific inference processors [75, 76] adopt DNN early stopping. It skips computations of the posterior layers if it judges them useless at the prior computation steps. Branch net-based inference prediction [75] uses layer-wise binary classifiers to determine whether remaining layers should be computed or not. Similarity maps generated in the Siamese network inform important spatial positions and it excludes computation of useless tiles in the remaining layers [76]. Another feature is the output ReLU prediction [27, 138] which predicts virtual output zeros by using only the minimum amount of pre-computation. It helps the sparsity exploitation core to skip both input and output sparsity during the inference scenario. However, these two features cannot be used during the training because the wrong prediction can disturb the training of the original backbone network. Instead, training processor utilizes features to compensate for drop in energy-efficiency during the inference. Run-time quantization [27, 29] and pruning automation [26] are unique features of training processors even though they are useless to inference processors.

8.8 Conclusion

The current AI system is smart only for the given applications, but what we expect for AI is not to simply provide the pre-determined routine processing. AI will cooperate together with human beings by actively thinking and acting. The purpose of the DNN training processor is to satisfy this desire. However, the design of the training processor is not that easy because it has more challenges compared with the inference processors. Training processors should support transpose-read of the weight during the EP stage. It needs to exploit three different types of sparsity: (1) input (2) weight, and (3) output sparsity. The new bit-precision optimization methods are also required to realize training at the micro-AI systems. As analyzed in Sect. 8.5, the design of the training processor should be accompanied by the algorithm–hardware co-optimization, and it has unique design methodologies compared with the conventional inference processors.

In spite of the many realizations of DNN training, there are still challenges. Firstly, current training processors showed few optimization methods during the WG stages. As explained in Sect. 8.4, primal weight needs high bit-precision to maintain its training performance. Moreover, it shows memory-intensive computation because of the elementwise multiplication and large kernel convolution. To release the memory burden of the WG stage, both a new algorithmic approximation and a hardware architecture are needed. The second issue is backward locking. Even if the acceleration of each layer becomes much faster, backward locking will eventually be the major obstacle to speed up training. Backward unlocking should be studied with the new algorithm to replace conventional BP based training.

The majority of training processors adopted long-term personalization or customization for their target applications by training with personal datasets. However, in the future AI research, online training will become mainstream. The authors in [29, 34] already showed that the DNN trained by general knowledge showed poor performance in an unexpected situation. Moreover, in the 6G wireless communication network, the need for online training will be essential because the pre-trained network cannot respond to the nonstationary nature of real-world situations to find optimal spectrum and base station [77]. Only online training can achieve such complicated real-time adaptation to realize the reliable 6G system. The application shift to online training will further highlight the necessity of an ultra-high-speed and ultra-low-power DNN training processor and system-level optimization.

Chapter 9
Conclusion

In this thesis, we introduced the overview of on-device training and our proposed on-device training SoC and systems. As explained in Chap. 1, on-device DNN training is needed due to evolution, advancement, and adaptation. According to the type of on-device scenario, the detailed requirements and conditions of the SoC design can be varied. New challenges of the mobile DNN training processor design are summarized as follows: (1) reading of transposed-weight during the error propagation (EP) stage, (2) high throughput even with the limited hardware resources, (3) higher bit-precision compared with the inference-only-supporting processors, (4) memory-intensive operations during weight-gradient-update (WG) stage, and (5) backward locking problem (Fig. 9.1).

In the software level, we proposed two direct feedback alignment (DFA) based DNN training algorithms. In Chap. 3, we firstly introduce the algorithm focusing on fine-tuning functionality [72]. It is advanced in Chap. 4 to be utilized for training from the scratch [74]. The initial version of new DFA algorithm just focused only on fully connected layer [72], but it is extended to general convolutional layers or recurrent path to support convolutional neural network or recurrent neural network [74]. Both algorithms solve the backward locking problem, and a new on-device training processor, DF-LNPU [34], is designed to optimize new DFA algorithm and details about it are introduced in Chap. 5.

© The Author(s), under exclusive license to Springer Nature Switzerland AG 2023

D. Han, H.-J. Yoo, *On-Chip Training NPU - Algorithm, Architecture and SoC Design*,
https://doi.org/10.1007/978-3-031-34237-0_9

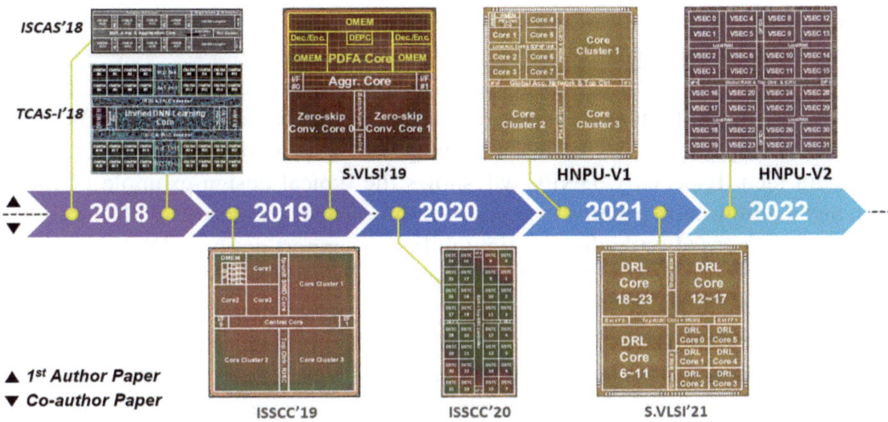

Fig. 9.1 On-chip DNN training processors which Donghyeon Han participated in

The DF-LNPU [34] utilizes the DFA algorithm but modifies it to pipelined DFA (PDFA) which constructs pipeline stages among three training stages to reduce latency caused by backward locking. It consists of feed-forward (FF), EP, and WG-optimized processors which show different core architecture to be optimized for different computational characteristics.

Not only DF-LNPU but also HNPU-V1 [29] and HNPU-V2 [80] are proposed for on-device training. As illustrated in Chap. 6, the HNPU-V1 adopts layer-wise adaptive precision scaling which finds the optimal bit-precision during the DNN training, and it utilizes bit-slice serial architecture to give high bit-reconfigurability. Since the acceleration with optimal bit-precision is more efficient than the pruning-aware DNN training, we utilize only bit-precision-optimized DNN training on HNPU-V1 design but exploit bit-slice-level sparsity for further throughput improvement. Unlike the conventional word-level sparsity exploitation, it is more robust to non-ReLU functions and EP stage which shows 0% input sparsity during the computation. Although the slice-level sparsity exploitation shows better performance during the non-ReLU-based DNN training or EP stage, fine-grained zero-slice skipping can require high hardware cost. To resolve this problem, HNPU-V1 utilizes FIFO based simple zero-slice skipping to minimize the required hardware cost. It can support not only input slice but also output slice to further improve the throughput.

In Chap. 7, we develop a new on-device DNN training processor which is the advanced version of HNPU-V1. It removes conventional layer-wise adaptive precision scaling logics and output slice skipping circuits but optimizes its performance by suggesting intrinsic true random number generator based stochastic rounding and combining iterative channel reordering based workload balancing and pruning-aware training. In addition, it increases the programmability of the reconfigurable accumulation network by using the learning task flag to support DFA based backward unlocking for various kinds of network and scenarios. Thanks to the dynamic learning task allocation, it achieves low-latency DNN training for real-time environmental adaptation.

In Chap. 8, we summarized some on-device DNN training processors and analyzed the recent trends of the research. Both HNPU-V1 and HNPU-V2 become a typical example of on-device DNN training SoC thanks to its great performance and main idea. Especially, HNPU-V1 shows the typical design example of long-term and medium-term training to support evolution and advancement DNN training scenarios. It focused on high reconfigurability to support not only DNN inference but also DNN training with high throughput. Especially, the HNPU-V1 pays attention to minimize the accuracy degradation while minimizing the required hardware cost through the quantization or weight pruning. On the contrary, the HNPU-V2 focused on ultra-fast DNN training even though it might lose the accuracy a little.

In conclusion, the DNN training processor for micro-AI systems establishes its own research area and it shows unique characteristics compared with the conventional inference processors. In order to improve the intelligence of the device, research on the training processor should be continued. The DNN training may not be the only answer for AI to reach human intelligence, but it will advance the future of AI and human coexistence.

References

1. Y.-H. Chen et al., Eyeriss: an energy-efficient reconfigurable accelerator for deep convolutional neural networks. IEEE J. Solid-State Circuits **52**(1), 127–138 (2017)
2. S. Choi, J. Lee, K. Lee, H. Yoo, A 9.02mW CNN-stereo-based real-time 3D hand-gesture recognition processor for smart mobile devices, in *2018 IEEE International Solid-State Circuits Conference (ISSCC)* (2018), pp. 220–222
3. D. Shin et al., DNPU: An 8.1TOPS/W reconfigurable CNN-RNN processor for general-purpose deep neural networks, in *IEEE Int. Solid-State Circuits Conf. (ISSCC) Dig. Tech. Papers* (2017), pp. 240–241
4. J. Lee et al., A 21mW low-power recurrent neural network accelerator with quantization tables for embedded deep learning applications, in *Proc. IEEE Asian Solid-State Circuits Conf. (A-SSCC)* (2017), pp. 237–240
5. K. Bong et al., A low-power convolutional neural network face recognition processor and a CIS integrated with always-on face detector. IEEE J. Solid-State Circuits **53**(1), 115–123 (2018)
6. K. Ueyoshi et al., QUEST: a 7.49TOPS multi-purpose log-quantized DNN inference engine stacked on 96 MB 3D SRAM using inductive coupling technology in 40 nm CMOS, in *Proc. IEEE Int. Solid-State Circuits Conf.* (2018), pp. 216–218
7. J. Lee et al., UNPU: a 50.6TOPS/W unified deep neural network accelerator with 1b-to-16b fully-variable weight bit-precision, in *2018 IEEE International Solid-State Circuits Conference (ISSCC)* (San Francisco, 2018), pp. 218–220
8. A. Amravati et al., A 55nm time-domain mixed-signal neuromorphic accelerator with stochastic synapses and embedded reinforcement learning for autonomous micro-robots, in *2018 IEEE International Solid-State Circuits Conference (ISSCC)* (2018), pp. 124–126
9. Z. Jiang et al., XNOR-SRAM: in-memory computing SRAM macro for binary/ternary deep neural networks, in *2018 IEEE Symposium on VLSI Technology* (2018), pp. 173–174
10. S. Kang et al., B-face: 0.2 MW CNN-based face recognition processor with face alignment for mobile user identification, in *2018 IEEE Symposium on VLSI Circuits* (Honolulu, 2018), pp. 137–138
11. J. Song et al., 7.1 an 11.5 TOPS/W 1024-MAC butterfly structure dual-core sparsity-aware neural processing unit in 8 nm flagship mobile SoC, in *2019 IEEE International Solid-State Circuits Conference (ISSCC)* (IEEE, 2019)
12. J. Lee et al., A full HD 60 fps CNN super resolution processor with selective caching based layer fusion for mobile devices, in *2019 Symposium on VLSI Circuits* (Kyoto, 2019), pp. C302–C303

© The Author(s), under exclusive license to Springer Nature Switzerland AG 2023
D. Han, H.-J. Yoo, *On-Chip Training NPU - Algorithm, Architecture and SoC Design*,
https://doi.org/10.1007/978-3-031-34237-0

13. J.-H. Kim et al., Z-PIM: an energy-efficient sparsity aware processing-in-memory architecture with fully-variable weight precision, in *2020 IEEE Symposium on VLSI Circuits* (2020)
14. Google, Tensor processing unit—second generation (TPU-v2)
15. NVIDIA, GTX 1080 Ti. https://www.geforce.com/hardware/desktop-gpus/geforce-gtx-1080-ti/specifications
16. NVIDIA, Tesla V100. https://www.nvidia.com/ko-kr/data-center/v100
17. B. Fleischer et al., A scalable multi- teraOPS deep learning processor core for ai training and inference, in *2018 IEEE Symposium on VLSI Circuits* (Honolulu, 2018), pp. 35–36
18. J. Oh et al., A 3.0 TFLOPS 0.62V scalable processor core for high compute utilization AI training and inference, *2020 IEEE Symposium on VLSI Circuits* (Honolulu, 2020), pp. 1–2
19. A. Agrawal et al., 9.1 A 7nm 4-core AI chip with 25.6TFLOPS hybrid FP8 training, 102.4TOPS INT4 inference and workload-aware throttling, in *2021 IEEE International Solid-State Circuits Conference (ISSCC)* (San Francisco, 2021), pp. 144–146
20. C. Kim et al., A 2.1TFLOPS/W mobile deep RL accelerator with transposable PE array and experience compression, in *2019 IEEE International Solid- State Circuits Conference (ISSCC)* (San Francisco, 2019), pp. 136–138
21. J. Lee et al., 7.7 LNPU: a 25.3TFLOPS/W sparse deep-neural-network learning processor with fine-grained mixed precision of FP8-FP16, in *2019 IEEE International Solid- State Circuits Conference (ISSCC)*
22. C. Lu et al., A 2.25 TOPS/W fully-integrated deep CNN learning processor with on-chip training, in *2019 IEEE Asian Solid-State Circuits Conference (A-SSCC)* (Macau, 2019), pp. 65–68
23. S. Yin et al., A 2.6 TOPS/W 16-bit fixed-point convolutional neural network learning processor in 65-nm CMOS. IEEE Solid-State Circuits Lett. **3**, 13–16 (2020)
24. S. Kang et al., 7.4 GANPU: a 135TFLOPS/W multi-DNN training processor for GANs with speculative dual-sparsity exploitation, in *2020 IEEE International Solid- State Circuits Conference (ISSCC)* (San Francisco, 2020), pp. 140–142
25. J. Su et al., 15.2 A 28 nm 64 Kb inference-training two-way transpose multibit 6T SRAM compute-in-memory macro for AI edge chips, in *2020 IEEE International Solid-State Circuits Conference (ISSCC)* (2020), pp. 240–242
26. S. Kim et al., A 146.52 TOPS/W deep-neural-network learning processor with stochastic coarse-fine pruning and adaptive input/output/weight skipping, in *2020 IEEE Symposium on VLSI Circuits* (Honolulu, 2020), pp. 1–2
27. F. Tu et al., Evolver: a deep learning processor with on-device quantization-voltage-frequency tuning, in *2021 IEEE Journal of Solid-State Circuits*
28. J. Park et al., 9.3 A 40 nm 4.81TFLOPS/W 8b floating-point training processor for non-sparse neural networks using shared exponent bias and 24-way fused multiply-add tree, in *2021 IEEE International Solid- State Circuits Conference (ISSCC)*
29. D. Han et al., HNPU: an adaptive DNN training processor utilizing stochastic dynamic fixed-point and active bit-precision searching, in *IEEE Journal of Solid-State Circuits*
30. J. Lee et al., OmniDRL: a 29.3 TFLOPS/W deep reinforcement learning processor with dualmode weight compression and on-chip sparse weight transposer, in *2021 Symposium on VLSI Circuits* (2021), pp. 1–2
31. Y. Wang et al., A 28 nm 276.55TFLOPS/W sparse deep-neural-network training processor with implicit redundancy speculation and batch normalization reformulation, in *2021 Symposium on VLSI Circuits* (2021), pp. 1–2
32. S. Kim et al., An energy-efficient GAN accelerator with on-chip training for domain-specific optimization, in *IEEE Journal of Solid-State Circuits*
33. Z. Yuan et al., Sticker: a 0.41-62.1 TOPS/W 8bit neural network processor with multi-sparsity compatible convolution arrays and online tuning acceleration for fully connected layers, in *Proc. IEEE Symp. VLSI Circuits* (2018), pp. 33–34
34. D. Han et al., DF-LNPU: a pipelined direct feedback alignment-based deep neural network learning processor for fast online learning. IEEE JSSC **56**(5), 1630–1640

35. S. Choi et al., An energy-efficient deep convolutional neural network training accelerator for in situ personalization on smart devices. IEEE J. Solid-State Circuits **55**(10), 2691–2702 (2020)
36. V. Sze et al., Efficient processing of deep neural networks: a tutorial and survey. Proc. IEEE **105**(12), 2295–2329 (2017)
37. Y. Chen et al., A survey of accelerator architectures for deep neural networks. Engineering **6**(3), 264–274 (2020)
38. M. Capra et al., An updated survey of efficient hardware architectures for accelerating deep convolutional neural networks. Fut. Internet **12**(7), 113 (2020)
39. J. Lee et al., The hardware and algorithm co-design for energy-efficient DNN processor on edge/mobile devices. IEEE Trans. Circuits Syst. I: Regular Papers **67**(10), 3458–3470 (2020)
40. D.E. Rumelhart, G.E. Hinton, R.J. Williams, Learning internal representations by back-propagating errors. Nature **323**, 533–536 (1986)
41. H. Nam, B. Han, Learning multi-domain convolutional neural networks, in *Proceedings of Computer Vision and Pattern Recognition (CVPR)* (2016)
42. M. Farhadi et al., TKD: temporal knowledge distillation for active perception, in *2020 IEEE Winter Conference on Applications of Computer Vision (WACV)* (Snowmass Village, 2020), pp. 942–951
43. D. Han, J. Lee, J. Lee, S. Choi, H.-J. Yoo, A 141.4 mw low-power online deep neural network training processor for real-time object tracking in mobile devices, in *Proceedings of International Symposium on Circuits and Systems (ISCAS)* (2018)
44. T.P. Lillicrap, D. Cownden, D.B. Tweed, C.J. Akerman, Random feedback weights support learning in deep neural networks. Nat. Commun. **7** (2016)
45. S. Han et al., EIE: efficient inference engine on compressed deep neural network, in *ACM/IEEE 43rd Annual International Symposium on Computer Architecture (ISCA)* (2016)
46. S. Han et al., ESE: efficient speech recognition engine with sparse LSTM on FPGA, in *Proceedings of the 2017 ACM/SIGDA International Symposium on Field-Programmable Gate Arrays (FPGA '17)*, pp. 75–84
47. J. Zhang et al., Eager pruning: algorithm and architecture support for fast training of deep neural networks, in *2019 ACM/IEEE 46th Annual International Symposium on Computer Architecture (ISCA)* (Phoenix, 2019), pp. 292–303
48. P. Dai et al., SparseTrain: exploiting dataflow sparsity for efficient convolutional neural networks training, in *2020 57th ACM/IEEE Design Automation Conference (DAC)* (San Francisco, 2020), pp. 1–6
49. E. Qin et al., SIGMA: a sparse and irregular GEMM accelerator with flexible interconnects for DNN training, in *2020 IEEE International Symposium on High Performance Computer Architecture (HPCA)* (2020), pp. 58–70
50. D. Yang et al., Procrustes: a dataflow and accelerator for sparse deep neural network training, in *2020 53rd Annual IEEE/ACM International Symposium on Microarchitecture (MICRO)* (Athens, 2020), pp. 711–724
51. M. Mahmoud et al., TensorDash: exploiting sparsity to accelerate deep neural network training, in *2020 53rd Annual IEEE/ACM International Symposium on Microarchitecture (MICRO)* (2020), pp. 781–795
52. B. Xu, N. Wang, T. Chen, M. Li, Empirical evaluation of rectified activations in convolution network, in *ICML Deep Learning* (2015), pp. 1–5
53. D. Misra, Mish: A self regularized non-monotonic activation function (2019). arXiv:1908.08681
54. NVIDIA, A100 Tensor Core GPU Architecture. https://www.nvidia.com/content/dam/en-zz/Solutions/Data-Center/nvidia-ampere-architecture-whitepaper.pdf
55. U. Köster et al., Flexpoint: an adaptive numerical format for efficient training of deep neural networks, in *Proceedings of the 31st International Conference on Neural Information Processing Systems (NIPS'17)* (Curran Associates, Red Hook), pp. 1740–1750
56. N. Wang et al., Training deep neural networks with 8-bit floating point numbers, in *Proc. Adv. Neural Inf. Process. Syst.* (2018), pp. 7685–7694

57. X. Sun et al., Hybrid 8-bit floating point (HFP8) training and inference for deep neural networks, in *Proc. Adv. Neural Inf. Process. Syst.* (2019), pp. 4901–4910
58. X. Sun et al., Ultra-low precision 4-bit training of deep neural networks, in *Proc. Adv. Neural Inf. Process. Syst.*, vol. 33 (2020)
59. S. Gupta et al., Deep learning with limited numerical precision (2015), pp. 1–10. CoRR, vol. abs/1502.02551
60. F. Seide et al., 1-bit stochastic gradient descent and its application to data-parallel distributed training of speech DNNs. INTERSPEECH (2014)
61. D. Alistarh et al., QSGD: communication-efficient SGD via gradient quantization and encoding, in *Proceedings of the 31st International Conference on Neural Information Processing Systems (NIPS'17)*, pp. 1707–1718
62. W. Wen et al., TernGrad: ternary gradients to reduce communication in distributed deep learning, in *Proceedings of the 31st International Conference on Neural Information Processing Systems (NIPS'17)*, pp. 1508–1518
63. Y. Lin et al., Deep gradient compression: reducing the communication bandwidth for distributed training (2017). CoRR, vol. abs/1712.01887
64. N. Strom et al., Scalable distributed DNN training using commodity GPU cloud computing, in *Sixteenth Annual Conference of the International Speech Communication Association* (2015)
65. A. Fikri Aji et al., Sparse communication for distributed gradient descent. In Empirical Methods, in *Natural Language Processing (EMNLP)* (2017)
66. D. Shin et al., Prediction confidence based low complexity gradient computation for accelerating DNN training, in *Proceedings of the 57th ACM/EDAC/IEEE Design Automation Conference (DAC '20)* (IEEE Press), Article 168, pp. 1–6
67. J. Shin et al., A pragmatic approach to on-device incremental learning system with selective weight updates, in *2020 57th ACM/IEEE Design Automation Conference (DAC)* (San Francisco, 2020), pp. 1–6
68. J. Li et al., TNPU: an efficient accelerator architecture for training convolutional neural networks, in *Proceedings of the 24th Asia and South Pacific Design Automation Conference (ASPDAC '19)*, pp. 450–455
69. Z. Huo et al., Decoupled Parallel Backpropagation with Convergence Guarantee (ICML, 2018)
70. Z. Huiping et al., Fully decoupled neural network learning using delayed gradients, in *IEEE Transactions on Neural Networks and Learning Systems* (2021)
71. A. Nokland, Direct feedback alignment provides learning in deep neural networks, in *Proceedings of Neural Information Processing Systems (NIPS)* (2016)
72. D. Han, H.-J. Yoo, Efficient convolutional neural network training with direct feedback alignment (2019). arXiv:1901.01986
73. B. Crafton, A. Parihar, E. Gebhardt, A. Raychowdhury, Direct feedback alignment with sparse connections for local learning (2019). arXiv:1903.02083
74. D. Han et al., Extension of direct feedback alignment to convolutional and recurrent neural network for bio-plausible deep learning (2020). arXiv: 2006.12830. https://arxiv.org/abs/2006.12830
75. Y. Kim et al., A 0.22–0.89 mW low-power and highly-secure always-on face recognition processor with adversarial attack prevention. IEEE Trans. Circuits Syst. II: Exp. Briefs **67**(5), 846–850 (2020)
76. S. Kim et al., A 64.1 mW accurate real-time visual object tracking processor with spatial early stopping on siamese network, in *IEEE Transactions on Circuits and Systems II: Express Briefs*
77. C. She et al., A tutorial on ultrareliable and low-latency communications in 6G: integrating domain knowledge into deep learning. Proc. IEEE **109**(3), 204–246 (2021)
78. Y. Wang et al., SWPU: a 126.04 TFLOPS/W edge-device sparse DNN training processor with dynamic sub-structured weight pruning, in *IEEE Transactions on Circuits and Systems I: Regular Papers* (2022)

79. S. Song et al., A 49.5 mW multi-scale linear quantized online learning processor for real-time adaptive object detection. IEEE Trans. Circuits Syst. II: Exp. Briefs **69**(5), 2443–2447 (2022)
80. D. Han et al., A 0.95 mJ/frame DNN training processor for robust object detection with real-world environmental adaptation, *2022 IEEE 4th International Conference on Artificial Intelligence Circuits and Systems (AICAS)* (2022), pp. 37–40
81. J. Deng, W. Dong, R. Socher, L. Li, K. Li, L. Fei-Fei, ImageNet: a large-scale hierarchical image database, in *2009 IEEE Conference on Computer Vision and Pattern Recognition* (Miami, 2009), pp. 248–255
82. J. Konecný et al., Federated learning: Strategies for improving communication efficiency, in *NIPS Workshop on Private Multi-Party Machine Learning* (2016)
83. H. Cai et al., TinyTL: reduce memory, not parameters for efficient on-device learning, in *Proceedings of the 34th International Conference on Neural Information Processing Systems (NIPS'20)* (2020)
84. J. Lin et al., On-device training under 256 KB memory, in *Proceedings of the 36th International Conference on Neural Information Processing Systems (NIPS'22)* (2022)
85. S.P. Singh, A. Kumar, H. Darbari, L. Singh, A. Rastogi, S. Jain, Machine translation using deep learning: an overview. International Conference on Computer, in *Communications and Electronics (Comptelix)* (2017), pp. 162–167
86. Y. Wang, R. Skerry-Ryan, D. Stanton, Y. Wu, R.J. Weiss, N. Jaitly, Z. Yang, Y. Xiao, Z. Chen, S. Bengio, Tacotron: Towards end-to-end speech synthesis (Interspeech, 2017)
87. W. Rawat, Z. Wang, Deep convolutional neural networks for image classification: a comprehensive review Neural Comput. **29**(9) (2017)
88. Y. LeCun, L. Bottou, Y. Bengio, P. Haffner, Gradient-based learning applied to document recognition. Proc. IEEE **86**, 2278–2324 (1998)
89. J. Hopfield, Neural networks and physical systems with emergent collective computational abilities. Proc. Natl. Acad. Sci. USA **79**, 2554–2558 (1982)
90. T. Salimans, J. Ho, X. Chen, S. Sidor, I. Sutskever, Evolution strategies as a scalable alternative to reinforcement learning (2017). ArXiv
91. T.P. Lillicrap, D. Cownden, D.B. Tweed, C.J. Akerman, Random feedback weights support learning in deep neural networks (2014). arXiv:1411.0247
92. Q. Liao, J.Z. Leibo, T. Poggio, How important is weight symmetry in backpropagation? (2015). arXiv:1510.05067
93. A. Krizhevsky, I. Sutskever, G.E. Hinton, ImageNet classification with deep convolutional neural networks, in *Proceedings of Neural Information Processing Systems (NIPS)* (2012), pp. 1106–1114
94. K. Simonyan, A. Zisserman, Very deep convolutional networks for large-scale image recognition, in *Proceedings of Computer Vision and Pattern Recognition (CVPR)* (2015)
95. V. Nair, G.E. Hinton, Rectified linear units improve restricted Boltzmann machines, in *Proceedings of International Conference on Machine Learning (ICML)* (2010)
96. I.J. Goodfellow, J. Pouget-Abadie, M. Mirza, B. Xu, D. Warde-Farley, S. Ozair, A. Courville, Y. Bengio, Generative adversarial nets, in *Proceedings of Neural Information Processing Systems (NIPS)* (2014)
97. B. Xu, N. Wang, T. Chen, M. Li, Empirical evaluation of rectified activations in convolutional network (2015). arXiv:1505.00853
98. N. Srivastava, G.E. Hinton, A. Krizhevsky, I. Sutskever, R. Salakhutdinov, Dropout: a simple way to prevent neural networks from overfitting. J. Mach. Learn. Res. **15**(1), 1929–1958 (2014)
99. A. Krizhevsky, Learning multiple layers of features from tiny images (2009)
100. I. Sutskever, J. Martens, G. Dahl, G.E. Hinton, On the importance of initialization and momentum in deep learning, in *Proceedings of International Conference on Machine Learning (ICML)* (2013), pp. 1139–1147
101. K. He, X. Zhang, S. Ren, J. Sun, Delving deep into rectifiers: Surpassing human-level performance on ImageNet classification (2015). CoRR, abs/1502.01852. http://arxiv.org/abs/1502.01852

102. E. Hoffer, I. Hubara, D. Soudry, Fix your classifier: the marginal value of training the last weight layer, in *International Conference on Learning Representations* (2018)
103. B. Han, J. Sim, H. Adam, BranchOut: Regularization for online ensemble tracking with convolutional neural networks, in *Proceedings of Computer Vision and Pattern Recognition (CVPR)* (2017)
104. S. Yun, J. Choi, Y. Yoo, K. Yun, J.Y. Choi, Action-decision networks for visual tracking with deep reinforcement learning, in *Proceedings of Computer Vision and Pattern Recognition (CVPR)* (2017)
105. M. Kristan, R. Pflugfelder, A. Leonardis, J. Matas, F. Porikli, L. Cehovin, G. Nebehay, G. Fernandez, T. Vojir et al., The visual object tracking VOT2013 challenge results, in *International Conference on Computer Vision Workshops (ICCVW)* (2013)
106. M. Kristan, R. Pflugfelder, A. Leonardis, J. Matas, L. Cehovin, G. Nebehay, G. Fernandez et al., The visual object tracking VOT2014 challenge results, in *European Conference on Computer Vision Workshops (ECCVW)* (2016)
107. Y. Wu, J. Lim, M. Yang, Object tracking benchmark. Trans. Pattern Anal. Mach. Intell. **37**, 1834–1848 (2015)
108. S. Hochreiter, J. Schmidhubers, Long short-term memory. Neural Comput. **9**, 1735–1780 (1997)
109. J. Chung, C. Gulcehre, K. Cho, Y. Bengio, Empirical evaluation of gated recurrent neural networks on sequence modeling (2014). ArXiv:1412.3555
110. S. Grossberg, Competitive learning: from interactive activation to adaptive resonance. Cogn. Sci. **11**(1), 23–63 (1987)
111. M. Jaderberg, W.M. Czarnecki, S. Osindero, O. Vinyals, A. Graves, D. Silver, K. Kavukcuoglu, Decoupled neural interfaces using synthetic gradients, in *Proceedings of the International Conference on Machine Learning (ICML)*, vol. 70 (2017), pp. 1627–1635
112. D.H. Lee, S. Zhang, A. Fischer, Y. Bengio, Difference target propagation, in *European Conference on Machine Learning and Principles and Practice of Knowledge Discovery in Databases (ECML-PKDD)*, vol. 9284 (2015), pp. 498–515
113. X. Zhang, X. Zhou, M. Lin, J. Sun, ShuffleNet: an extremely efficient convolutional neural network for mobile devices (2017). arXiv:1707.01083
114. D.P. Kingma, J. Ba, Adam: a method for stochastic optimization (2014). arXiv:1412.6980
115. R.J. Williams, D. Zipser, *Backpropagation: Theory, Architectures and Applications* (Erlbaum, Hillsdale, 1992), pp. 433–486
116. K. He, X. Zhang, S. Ren, J. Sun, Deep residual learning for image recognition (2015). arXiv:1512.03385
117. F. Yu, V. Koltun, Multi-scale context aggregation by dilated convolutions, in *International Conference on Learning Representations (ICLR)* (2016)
118. D. Im, D. Han, S. Choi, S. Kang, H.-J. Yoo, DT-CNN: dilated and transposed convolution neural network accelerator for real-time image segmentation on mobile devices, in *Proceedings of International Symposium on Circuits and Systems (ISCAS)* (2019)
119. S. Merity, C. Xiong, J. Bradbury, R. Socher, Pointer sentinel mixture models (2016). arXiv:1609.07843
120. D. Han, J. Lee, J. Lee, H.-J. Yoo, A 1.32 TOPS/W energy efficient deep neural network learning processor with direct feedback alignment based heterogeneous core architecture, in *Symposium on VLSI Circuits* (2019)
121. H.B. McMahan, D. Ramage, Federated learning: collaborative machine learning without centralized training data (2017). https://research.googleblog.com/2017/04/federated-learning-collaborative.html
122. A. Wang, A. Singh, J. Michael, F. Hill, O. Levy, S.R. Bowman, GLUE: a multi-task benchmark and analysis platform for natural language understanding, in *Proceedings of the International Conference on Learning Representations (ICLR)* (2018)
123. D. Han, J. Lee, J. Lee, H.-J. Yoo, A low-power deep neural network online learning processor for real-time object tracking application. IEEE Trans. Circuits Syst. I: Regular Papers **66**(5) (2019)

124. S. Hong, T. You, S. Kwak, B. Han, Online tracking by learning discriminative saliency map with convolutional neural network. https://arxiv.org/abs/1502.06796
125. N. Wang, S. Li, A. Gupta, D.-Y. Yeung, Transferring rich feature hierarchies for robust visual tracking. https://arxiv.org/abs/1501.04587
126. A. Milan, S.H. Rezatofighi, A. Dick, I. Reid, K. Schindler, Online multi-target tracking using recurrent neural networks
127. M. Horowitz, 1.1 computing's energy problem (and what we can do about it), in *IEEE Int. Solid-State Circuits Conf. (ISSCC) Dig. Tech. Papers* (2014), pp. 10–14
128. D. Kim, J. Kung, S. Mukhopadhyay, Power-aware digital multilayer perceptron accelerator with on-chip training based on approximate computing, in *IEEE Trans. Emerging Topics in Computing (TETC)* (2017)
129. Z. Huo, B. Gu, H. Huang, Training neural networks using features replay, in *Proc. Adv. Neural Inf. Process. Syst.* (2018), pp. 6660–6669
130. S. Vigna, Further scramblings of Marsaglia's xorshift generators. J. Comput. Appl. Math. **315**, 175–181 (2016)
131. I. Jung, J. Son, M. Baek, B. Han, Real-time MDNet, in *Proceedings of the European Conference on Computer Vision (ECCV)*, pp. 83–98
132. Q. Wang, L. Zhang, L. Bertinetto, W. Hu, P.H. Torr, Fast online object tracking and segmentation: a unifying approach. arXiv preprint arXiv:1812.05050
133. A. Loquercio et al., DroNet: learning to fly by driving. IEEE Robot. Autom. Lett. **3**(2), 1088–1095 (2018)
134. S. Du, H. Guo, A. Simpson, Self-driving car steering angle prediction based on image recognition (2019). arXiv preprint arXiv:1912.05440
135. S. Zhou et al., DoReFa-Net: training low bitwidth convolutional neural networks with low bitwidth gradients. arXiv preprint arXiv:1606.06160
136. X. Chen et al., FxpNet: training a deep convolutional neural network in fixed-point representation, in *2017 International Joint Conference on Neural Networks (IJCNN)* (Anchorage, 2017), pp. 2494–2501
137. A. HeydariGorji et al., HyperTune: dynamic hyperparameter tuning for efficient distribution of DNN training over heterogeneous systems, in *Proceedings of the 39th International Conference on Computer-Aided Design (ICCAD)*
138. S. Kang et al., GANPU: an energy-efficient multi-DNN training processor for GANs with speculative dual-sparsity exploitation, in *2021 IEEE Journal of Solid-State Circuits*
139. D. Wang et al., signADAM++: learning confidences for deep neural networks, in *2019 International Conference on Data Mining Workshops (ICDMW)*
140. Y. Wang et al., HPPU: an energy-efficient sparse DNN training processor with hybrid weight pruning, in *2021 IEEE 3rd International Conference on Artificial Intelligence Circuits and Systems (AICAS)*
141. Y. Gong et al., RAODAT: an energy-efficient reconfigurable AI-based object detection and tracking processor with online learning, in *Proc. IEEE Asian Solid-State Circuits Conf. (A-SSCC)*, pp. 1–3
142. Y. He et al., Filter pruning via geometric median for deep convolutional neural networks acceleration, in *Proceedings of Computer Vision and Pattern Recognition (CVPR)* (2019), pp. 4335–4344
143. L. Bassham et al., A statistical test suite for random and pseudorandom number generators for cryptographic applications, NIST SP
144. A. Nagabandi et al., Learning to adapt in dynamic, real-world environments through meta-reinforcement learning. arXiv preprint arXiv:1803.11347
145. D. Han et al., A mobile DNN training processor with automatic bit precision search and fine-grained sparsity exploitation. IEEE Micro **42**(2), 16–25 (2022)
146. M. Giordano et al., CHIMERA: A 0.92 TOPS, 2.2 TOPS/W edge AI accelerator with 2 MByte on-chip foundry resistive RAM for efficient training and inference, in *2021 Symposium on VLSI Circuits* (2021), pp. 1–2

147. G. Cybenko, Approximation by superpositions of a sigmoidal function. Math. Control Signals Syst. **2**(4), 303–314 (1989). https://doi.org/10.1007/BF02551274
148. Z. Li et al., Learning without forgetting, in *European Conference on Computer Vision (ECCV)* (2016)
149. J.H. Holland, *Adaptation in Natural and Artificial Systems* (University of Michigan Press, Ann Arbor)
150. G. Bellec et al., A solution to the learning dilemma for recurrent networks of spiking neurons. Nat. Commun. **11**, 3625 (2020)
151. S. Seung et al., A correlation game for unsupervised learning yields computational interpretations of Hebbian excitation, anti-Hebbian inhibition, and synapse elimination (2017). arXiv preprint arXiv:1704.00646
152. P. Baldi et al., A theory of local learning, the learning channel, and the optimality of backpropagation, in *Neural Networks* (2016)
153. O. Senn, A predictive coding approach to estimating the complexity of popular music drum patterns (2022)
154. H. Mostafa et al., Deep supervised learning using local errors. Front. Neurosci. **12**, 608 (2018)
155. A. Nøkland, Training neural networks with local error signals, in *International Conference on Machine Learning* (PMLR, 2019)
156. R. Takiyama et al., Error-correction learning of three layer neural networks based on linear homogeneous expressions. IEICE Trans. Fundam. Electron. Commun. Comput. Sci. **76**(4), 637–641 (1993)
157. D.H. Ackley et al., A learning algorithm for Boltzmann machines. Cogn. Sci. **9**(1), 147–169 (1985)
158. S. Boyd et al., Randomized gossip algorithms. IEEE Trans. Inf. Theory **52**(6), 2508–2530 (2006)
159. M. Blot et al., Gossip training for deep learning (2016). arXiv:1611.09726. https://arxiv.org/abs/1611.09726
160. J. Daily et al., GossipGraD: scalable deep learning using gossip communication based asynchronous gradient descent (2018). arXiv:1803.05880. https://arxiv.org/abs/1803.05880
161. Z. Zhou et al., Edge intelligence: paving the last mile of artificial intelligence with edge computing. Proc. IEEE **107**(8), 1738–1762 (2019)
162. D. Han, S. Kang, S. Kim, J. Lee, H.-J. Yoo, Energy-Efficient DNN Training Processors on Micro-AI Systems, in IEEE Open Journal of the Solid-State Circuits Society, **2**, 259–275 (2022). https://doi.org/10.1109/OJSSCS.2022.3219034

Curriculum Vitae

Name	:	Donghyeon Han
Date of birth	:	May 10, 1995
Address	:	Semiconductor System Laboratory #1233, E3-2 (Electrical Engineering Bldg.), KAIST 291, Daehak-ro, Yuseong-gu, Daejeon, Republic of Korea
E-mail	:	hdh4797@kaist.ac.kr

Major Research Experience

1. **Intelligent Computer Vision Algorithm and Related Systems**
    - Built C++ software framework for the ICP-PSO based hand tracking algorithm
    - Invented ICP-PSO based automatic hand size analyzer
    - Designed C++ fingertip and depth detection system with hand tracking
    - Built C++ software framework for the face alignment with Local Binary Features (LBFs)
    - Built C/C++ software framework for the Support Vector Machine (SVM) classifier of face recognition
    - Designed C++/C# face recognition interface in Mobile Platform
    - Speeded up robust features (SURF) based object recognition system
    - Deep neural network (DNN) online learning based object tracking system
    - Real-time online DNN training based fast object detection system
2. **Deep Learning Algorithm**
    - Invented DNN training algorithm: binary feedback alignment (ISCAS'18)
    - Invented DNN training algorithm: binary direct feedback alignment (ICCVW'19)
    - Exponent-only output zero prediction algorithm (ISSCC'20)

© The Author(s), under exclusive license to Springer Nature Switzerland AG 2023

D. Han, H.-J. Yoo, *On-Chip Training NPU - Algorithm, Architecture and SoC Design*,
https://doi.org/10.1007/978-3-031-34237-0

- Direct feedback alignment algorithm for CNN and RNN training (ArXiv'20)
- New data representation for low-bit-precision DNN training (JSSC'21)
- Adaptive bit-precision scaling algorithm for DNN training (JSSC'21)

3. **System-on-Chip Design**

- Designed binary feedback alignment based homogeneous DNN training core (TCAS-I'18)
- Designed integral-RLC and transpose-RLC decoder (TCAS-I'18)
- Designed dropout controller for clock-gating-based energy-saving (TCAS-I'18)
- Designed memory efficient batch normalization, softmax module (ISSCC'19, A-SSCC'20)
- Designed DNN training processor with input sparsity exploitation in LNPU (ISSCC'19)
- Designed heterogeneous DNN online learning core with intra- and inter-core pipeline (JSSC'20)
- Designed direct error propagation core with binary direct feedback alignment (JSSC'20)
- Designed DNN online learning processor w/ Direct Feedback Alignment (JSSC'20)
- Designed area- and energy-efficient output zero prediction unit in GANPU (ISSCC'20)
- Designed dual-sparsity-aware accelerator for DNN training in GANPU (ISSCC'20)
- Designed bit-slice serial computing based DNN acceleration core (JSSC'21, ISSCC'22)
- Designed bit-slice-level sparsity exploitation core (JSSC'21)
- Designed adaptive bandwidth reconfigurable accumulation network (JSSC'21)

4. **Embedded Platform Design**

- Developed Haar-like based face detection FPGA IP (ISSCC'17)
- Developed Bluetooth based wireless communication FPGA IP (ISSCC'17)
- Designed Bluetooth board for mobile face recognition system (ISSCC'17)
- Designed embedded board for online DNN training-based object detection system (JSSC'21)

Educations

2011. 3. – 2013. 2.	Daejeon Science High School, South Korea (2-year accelerated graduation)
2013. 3. – 2017. 2.	B.S. in Electrical Engineering, KAIST (National Full-Scholarship)

2017. 3. – 2019. 2.	M.S. in Electrical Engineering, KAIST (National Full-Scholarship)
2019. 3. – 2023. 2.	Ph.D. in Electrical Engineering, KAIST

Career

2013. 3. – 2017. 2.	Korea Foundation for Advanced Studies, Undergraduate Student Scholarship Program
2013. 3. – 2017. 2.	National Science & Technology Scholarship for Undergraduate Student (except 2014)
2014. 3.	Dean's List in Recognition of Outstanding Scholastic Achievement
2014. 6. – 2014. 11.	Team Leader of the Volunteer Corps in KAIST MIDAM Scholarship
2014. 12. – 2015. 6.	The Representative of the KAIST MIDAM Scholarship
2015. 9.	Award Certificate in Korea MIDAM Scholarship
2016. 2.	16' Korea Electric Power Corporation (KEPCO) Scholarship for Electric Engineering Student
2018. 9.	KAIST Electrical Engineering Best Teaching Assistant Award
2021. 7. – 2022. 7.	Laboratory Leader of Semiconductor System Laboratory
2022. 6.	Best Paper Award at 2022 IEEE International Conference on Artificial Intelligence Circuits and Systems (AICAS 2022)
2022. 6.	Best Live Demo Award at 2022 IEEE International Conference on Artificial Intelligence Circuits and Systems (AICAS 2022)

Publications

1. **Donghyeon Han**, Jinsu Lee, Jinmook Lee, and Hoi-Jun Yoo, "A Low-Power Deep Neural Network Online Learning Processor for Real-Time Object Tracking Application," *IEEE Transactions on Circuits and Systems I: Regular Papers*, Nov. 2018.
2. **Donghyeon Han**, Jinsu Lee, and Hoi-Jun Yoo, "DF-LNPU: A Pipelined Direct Feedback Alignment-Based Deep Neural Network Learning Processor for Fast Online Learning," *IEEE Journal of Solid-State Circuits*, Dec. 2020.
3. **Donghyeon Han**, Dongseok Im, Gwangtae Park, Youngwoo Kim, Seokchan Song, Juhyoung Lee, and Hoi-Jun Yoo, "HNPU: An Adaptive DNN Training Processor Utilizing Stochastic Dynamic Fixed-point and Active Bit-precision Searching," *IEEE Journal of Solid-State Circuits*, Sep. 2021.
4. **Donghyeon Han**, Dongseok Im, Gwangtae Park, Youngwoo Kim, Seokchan Song, Juhyoung Lee, and Hoi-Jun Yoo, "A Mobile DNN Training Processor with Automatic Bit-precision Search and Fine-grained Sparsity Exploitation," *IEEE Micro*, Mar. 2020.
5. **Donghyeon Han**, Jinsu Lee, Jinmook Lee, Sungpill Choi and Hoi-Jun Yoo, "A 141.4 mW Low-Power Online Deep Neural Network Training Processor for Real-time Object Tracking in Mobile Devices," *IEEE International Symposium on Circuits and Systems*, May. 2018.
6. **Donghyeon Han**, Jinsu Lee, Jinmook Lee, and Hoi-Jun Yoo, "A 1.32 TOPS/W Energy Efficient Deep Neural Network Learning Processor with Direct Feed-back Alignment Based Heterogeneous Core Architecture," *IEEE Symposium on VLSI Circuits*, June. 2019.
7. **Donghyeon Han**, and Hoi-Jun Yoo, "Direct Feedback Alignment Based Convolutional Neural Network Training for Low-Power Online Learning Processor," *Proceedings of the IEEE/CVF International Conference on Computer Vision Workshop*, Oct. 2019.
8. **Donghyeon Han**, Dongseok Im, Gwangtae Park, Youngwoo Kim, Seokchan Song, Juhyoung Lee, and Hoi-Jun Yoo, "An Energy-Efficient Deep Neural Network Training Processor with Bit-Slice-Level Reconfigurability and Sparsity Exploitation," *IEEE Symposium in Low-Power and High-Speed Chips (COOL CHIPS)*, Apr. 2021.
9. **Donghyeon Han**, Dongseok Im, Gwangtae Park, Youngwoo Kim, Seokchan Song, Juhyoung Lee, and Hoi-Jun Yoo, "A 0.95 mJ/frame DNN Training Processor for Robust Object Detection with Real-World Environmental Adaptation," *IEEE International Conference on Artificial Intelligence Circuits and Systems (AICAS)*, Jun. 2022.
10. **Donghyeon Han**, Gwangtae Park, Junha Ryu, and Hoi-Jun Yoo, "Extension of Direct Feedback Alignment to Convolutional and Recurrent Neural Network for Bio-plausible Deep Learning," *arXiv:2006.12830*, Jun. 2020.
11. **Donghyeon Han**, Dongseok Im, Gwangtae Park, Youngwoo Kim, Seokchan Song, Juhyoung Lee, and Hoi-Jun Yoo, "HNPU-V2: A 46.6 FPS DNN Training

Processor for Real-World Environmental Adaptation Based Robust Object Detection on Mobile Devices," 2022 IEEE Symposium on High Performance Chips (HOT CHIPS), Aug. 2022.

12. Kyeongryeol Bong, Sungpill Choi, Changhyeon Kim, **Donghyeon Han**, and Hoi-Jun Yoo, "A Low-power Convolutional Neural Network Face Recognition Processor and a CIS Integrated with Always-on Face Detector," *IEEE Journal of Solid-State Circuits*, Jan. 2018.
13. Jinsu Lee, Juhyoung Lee, **Donghyeon Han**, Jinmook Lee, Gwantae Park, and Hoi-Jun Yoo, "An Energy-Efficient Sparse Deep-Neural-Network Learning Accelerator with Fine-Grained Mixed Precision of FP8-FP16," *IEEE Solid-State Circuits Letters*, Aug. 2019.
14. Sungpill Choi, Kyeongryeol Bong, **Donghyeon Han**, and Hoi-Jun Yoo, "CNNP-v2: A Memory-Centric Architecture for Low-Power CNN Processor on Domain-Specific Mobile Devices," *IEEE Journal on Emerging and Selected Topics in Circuits and Systems*, Nov. 2019.
15. Youngwoo Kim, **Donghyeon Han**, Changhyeon Kim, and Hoi-Jun Yoo, "A 0.22–0.89 mW Low-Power and Highly-Secure Always-on Face Recognition Processor with Adversarial Attack Prevention," *IEEE Transactions on Circuits and Systems II: Express Briefs*, Mar. 2020.
16. Dongseok Im, **Donghyeon Han**, Sungpill Choi, Sanghoon Kang, and Hoi-Jun Yoo, "DT-CNN: An Energy-Efficient Dilated and Transposed Convolutional Neural Network Processor for Region of Interest Based Image Segmentation," *IEEE Transactions on Circuits and Systems I: Regular Papers*, Mar. 2020.
17. Gwangtae Park, Dongseok Im, **Donghyeon Han**, and Hoi-Jun Yoo, "A 1.15 TOPS/W Energy-Efficient Capsule Network Accelerator for Real-Time 3D Point Cloud Segmentation in Mobile Environment," *IEEE Transactions on Circuits and Systems II: Express Briefs*, Jun. 2020.
18. Jinsu Lee, Sanghoon Kang, Jinmook Lee, Dongjoo Shin, **Donghyeon Han**, and Hoi-Jun Yoo, "The Hardware and Algorithm Co-Design for Energy-Efficient DNN Processor on Edge/Mobile Devices," *IEEE Transactions on Circuits and Systems I: Regular Papers*, Sep. 2020.
19. Kyoung-Rog Lee, Jihoon Kim, Changhyeon Kim, **Donghyeon Han**, Juhyoung Lee, Jinsu Lee, Hongsik Jeong, and Hoi-Jun Yoo, "A 1.02-μW STT-MRAM-Based DNN ECG Arrhythmia Monitoring SoC With Leakage-Based Delay MAC Unit," *IEEE Solid-State Circuits Letters*, Sep. 2020.
20. Soyeon Kim, Sangjin Kim, Sangyeob Kim, **Donghyeon Han**, and Hoi-Jun Yoo, "A 64.1mW Accurate Real-time Visual Object Tracking Processor with Spatial Early Stopping on Siamese Network," *IEEE Transactions on Circuits and Systems II: Express Briefs*, Mar. 2021.
21. Sanghoon Kang, **Donghyeon Han**, Juhyoung Lee, Dongseok Im, Sangyeob Kim, Soyeon Kim, Junha Ryu, and Hoi-Jun Yoo, "GANPU: An Energy-Efficient Multi-DNN Training Processor for GANs with Speculative Dual-Sparsity Exploitation," *IEEE Journal of Solid-State Circuits*, Sep. 2021.

22. Dongseok Im, **Donghyeon Han**, Sanghoon Kang, and Hoi-Jun Yoo, "A Pipelined Point Cloud Based Neural Network Processor for 3D Vision with Large-Scale Max Pooling Layer Prediction," *IEEE Journal of Solid-State Circuits*, Jun. 2021.
23. Soyeon Kim, Sanghoon Kang, **Donghyeon Han**, Sangjin Kim, Sangyeob Kim, and Hoi-Jun Yoo, "An Energy-Efficient GAN Accelerator with On-Chip Training for Domain Specific Optimization," *IEEE Journal of Solid-State Circuits*, Jul. 2021.
24. Sanghoon Kang, Gwangtae Park, Sangjin Kim, Soyeon Kim, **Donghyeon Han**, and Hoi-Jun Yoo, "An Overview of Sparsity Exploitation in CNNs for On-Device Intelligence with Software-Hardware Cross-Layer Optimizations," *IEEE Journal on Emerging and Selected Topics in Circuits and Systems*, Oct. 2021.
25. Sangyeob Kim, Juhyoung Lee, Sanghoon Kang, **Donghyeon Han**, Wooyoung Jo, and Hoi-Jun Yoo, "TSUNAMI: Triple Sparsity-aware Ultra Energy-efficient Neural Network Training Accelerator with Multi-modal Iterative Pruning," *IEEE Transactions on Circuits and Systems I: Regular Papers*, Jan. 2022.
26. Juhyoung Lee, Sangyeob Kim, Sangjin Kim, Wooyoung Jo, **Donghyeon Han**, and Hoi-Jun Yoo, "OmniDRL: An Energy-Efficient Deep Reinforcement Learning Processor with Dual-mode Weight Compression and Sparse Weight Transposer," *IEEE Journal of Solid-State Circuits*, Jan. 2022.
27. Sangwoo Ha, Sangjin Kim, **Donghyeon Han**, Soyeon Um, and Hoi-Jun Yoo, "A 36.2 dB High SNR and PVT/Leakage-robust eDRAM Computing-In-Memory Macro with Segmented BL and Reference Cell Array," *IEEE Transactions on Circuits and Systems II: Express Briefs*, Mar. 2022.
28. Seokchan Song, Soyeon Kim, Gwangtae Park, **Donghyeon Han**, and Hoi-Jun Yoo, "A 49.5 mW Multi-scale Linear Quantized Online Learning Processor for Real-Time Adaptive Object Detection," *IEEE Transactions on Circuits and Systems II: Express Briefs*, Mar. 2022.
29. Jinsu Lee, Juhyoung Lee, **Donghyeon Han**, and Hoi-Jun Yoo, "LNPU: A 25.3TFLOPS/W Sparse Deep-Neural-Network Learning Processor with Fine-Grained Mixed Precision of FP8-FP16," *IEEE International Solid-State Circuits Conference*, Feb. 2019.
30. Sungpil Choi, Kyeongryeol Bong, **Donghyeon Han**, and Hoi-Jun Yoo, "CNNP-v2: An Energy Efficient Memory-Centric Convolutional Neural Network Processor Architecture," *IEEE International Conference on Artificial Intelligence Circuits and Systems*, Mar. 2019.
31. Dongseok Im, **Donghyeon Han**, Sungpil Choi, Sanghoon Kang, and Hoi-Jun Yoo, "DT-CNN: Dilated and Transposed Convolution Neural Network Accelerator for Real-time Image Segmentation on Mobile Devices," *IEEE International Symposium on Circuits and Systems*, May. 2019.
32. Jinsu Lee, Juhyoung Lee, **Donghyeon Han**, and Hoi-Jun Yoo, "LNPU: An Energy-Efficient Deep-Neural-Network Training Processor with Fine-Grained

Mixed Precision," *IEEE Hot Chips: A Symposium on High Performance Chips*, Aug. 2019.

33. Sanghoon Kang, **Donghyeon Han**, Juhyoung Lee, Dongseok Im, Sangyeob Kim, Soyeon Kim, and Hoi-Jun Yoo, "7.4 GANPU: A 135TFLOPS/W multi-DNN training processor for GANs with speculative dual-sparsity exploitation," *IEEE International Solid-State Circuits Conference*, Feb. 2020.
34. Youngwoo Kim, **Donghyeon Han**, Changhyeon Kim, and Hoi-Jun Yoo, "A 0.22–0.89 mW Low-Power and Highly-Secure Always-on Face Recognition Processor with Adversarial Attack Prevention," *IEEE International Symposium on Circuits and Systems*, May. 2020.
35. Dongseok Im, Sanghoon Kang, **Donghyeon Han**, Sungpill Choi, Hoi-Jun Yoo, "A 4.45 ms Low-latency 3D Point-cloud-based Neural Network Processor for Hand Pose Estimation in Immersive Wearable Devices," *IEEE Symposium on VLSI Circuits*, June. 2020.
36. Gwangtae Park, Dongseok Im, **Donghyeon Han**, and Hoi-Jun Yoo, "A 1.15 TOPS/W Energy-Efficient Capsule Network Accelerator for Real-Time 3D Point Cloud Segmentation in Mobile Environment," *IEEE Symposium on Integrated Circuits and Systems*, Aug. 2020.
37. Sanghoon Kang, **Donghyeon Han**, Juhyoung Lee, Dongseok Im, Sangyeob Kim, Soyeon Kim, Junha Ryu, and Hoi-Jun Yoo, "GANPU: A Versatile Many-Core Processor for Training GAN on Mobile Devices with Speculative Dual-Sparsity Exploitation," *IEEE Hot Chips: A Symposium on High Performance Chips*, Aug. 2020.
38. Kyoung-Rog Lee, Jihoon Kim, Changhyeon Kim, **Donghyeon Han**, Juhyoung Lee, Jinsu Lee, Hongsik Jeong, and Hoi-Jun Yoo, "A 1.02-μW STT-MRAM-Based DNN ECG Arrhythmia Monitoring SoC With Leakage-Based Delay MAC Unit," *IEEE Asian Solid-State Circuits Conference (A-SSCC)*, Nov. 2020.
39. Soyeon Kim, Sanghoon Kang, **Donghyeon Han**, Sangyeob Kim, Sangjin Kim, and Hoi-jun Yoo, "An Energy-Efficient GAN Accelerator with On-chip Training for Domain Specific Optimization," *IEEE Asian Solid-State Circuits Conference (A-SSCC)*, Nov. 2020.
40. Soyeon Kim, Sangjin Kim, Sangyeob Kim, **Donghyeon Han**, and Hoi-Jun Yoo, "A 64.1mW Accurate Real-time Visual Object Tracking Processor with Spatial Early Stopping on Siamese Network," *IEEE International Symposium on Circuits and Systems*, May. 2021.
41. Juhyoung Lee, Sangyeob Kim, Sangjin Kim, Wooyoung Jo, **Donghyeon Han**, Jinsu Lee, and Hoi-Jun Yoo, "OmniDRL: A 29.3 TFLOPS/W Deep Reinforcement Learning Processor with Dual-mode Weight Compression and On-chip Sparse Weight Transposer," *IEEE Symposium on VLSI Circuits*, June. 2021.
42. Juhyoung Lee, Changhyeon Kim, **Donghyeon Han**, Sangyeob Kim, Sangjin Kim, and Hoi-Jun Yoo, "Energy-Efficient Deep Reinforcement Learning Accelerator Designs for Mobile Autonomous Systems," *IEEE International Conference on Artificial Intelligence Circuits and Systems*, Mar. 2019.

43. Juhyoung Lee, Jihoon Kim, Wooyoung Jo, Sangyeob Kim, Sangjin Kim, **Donghyeon Han**, Jinsu Lee, and Hoi-jun Yoo, "An Energy-efficient Floating-Point DNN Processor using Heterogeneous Computing Architecture with Exponent-Computing-in-Memory," *IEEE Hot Chips: A Symposium on High Performance Chips*, Aug. 2021.
44. Juhyoung Lee, Sangyeob Kim, Jihoon Kim, Sangjin Kim, Wooyoung Jo, **Donghyeon Han**, and Hoi-Jun Yoo, "OmniDRL: An Energy-Efficient Mobile Deep Reinforcement Learning Accelerators with Dual-mode Weight Compression and Direct Processing of Compressed Data," *IEEE Hot Chips: A Symposium on High Performance Chips*, Aug. 2021.
45. Surin Gweon, Sanghoon Kang, **Donghyeon Han**, Kyoung-Rog Lee, Kwantae Kim, and Hoi-Jun Yoo, "FlashMAC: An Energy-Efficient Analog-Digital Hybrid MAC with Variable Latency-Aware Scheduling," *IEEE Asian Solid-State Circuits Conference (A-SSCC)*, Nov. 2021.
46. Dongseok Im, Gwangtae Park, Zhiyong Li, Junha Ryu, Sanghoon Kang, **Donghyeon Han**, Jinsu Lee, and Hoi-Jun Yoo, "DSPU: A 281.6mW Real-Time Depth Signal Processing Unit for Deep Learning- Based Dense RGB-D Data Acquisition with Depth Fusion and 3D Bounding Box Extraction in Mobile Platforms," *IEEE International Solid-State Circuits Conference*, Feb. 2022.
47. Dongseok Im, Gwangtae Park, Junha Ryu, Zhiyong Li, Sanghoon Kang, **Donghyeon Han**, Jinsu Lee, Wonhoon Park and Hoi-Jun Yoo, "A Low-power and Real-time 3D Object Recognition Processor with Dense RGB-D Data Acquisition in Mobile Platforms," *IEEE Symposium in Low-Power and High-Speed Chips (COOL CHIPS)*, Apr. 2022.
48. Zhiyong Li, Sangjin Kim, Dongseok Im, **Donghyeon Han**, and Hoi-Jun Yoo, "An 0.92 mJ/frame High-quality FHD Super-resolution Mobile Accelerator SoC with Hybrid-precision and Energy-efficient Cache," *IEEE Custom Integrated Circuits Conference (CICC)*, Apr. 2022.
49. Sangwoo Ha, Sangjin Kim, **Donghyeon Han**, Soyeon Um, and Hoi-Jun Yoo, "A 36.2 dB High SNR and PVT/Leakage-robust eDRAM Computing-In-Memory Macro with Segmented BL and Reference Cell Array," *IEEE International Symposium on Circuits and Systems*, May. 2022.
50. Seokchan Song, Soyeon Kim, Gwangtae Park, **Donghyeon Han**, and Hoi-Jun Yoo, "A 49.5 mW Multi-Scale Linear Quantized Online Learning Processor for Real-Time Adaptive Object Detection," *IEEE International Symposium on Circuits and Systems*, May. 2022.
51. Dongseok Im, Gwangtae Park, Junha Ryu, Zhiyong Li, Sanghoon Kang, **Donghyeon Han**, Jinsu Lee, Wonhoon Park, and Hoi-Jun Yoo, "DSPU: A 281.6mW Real-Time Deep Learning-Based Dense RGB-D Data Acquisition with Sensor Fusion and 3D Perception System-on-Chip," 2022 IEEE Symposium on High Performance Chips (HOT CHIPS), Aug. 2022.
52. Zhiyong Li, Sangjin Kim, Dongseok Im, **Donghyeon Han**, and Hoi-Jun Yoo, "An Energy-efficient High-quality FHD Super-resolution Mobile Accelerator

SoC with Hybrid-precision and Energy-efficient Cache Subsystem," 2022 IEEE Symposium on High Performance Chips (HOT CHIPS), Aug. 2022.

53. Sangjin Kim, Zhiyong Li, Soyeon Um, Wooyoung Jo, Sangwoo Ha, Juhyoung Lee, Sangyeob Kim, **Donghyeon Han**, and Hoi-Jun Yoo, "DynaPlasia: An eDRAM In-Memory-Computing-based Reconfigurable Spatial Accelerator with Triple-mode Cell for Dynamic Resource Switching," *IEEE International Solid-State Circuits Conference*, Feb. 2023.
54. Sangyeob Kim, Soyeon Kim, Seongyon Hong, Sangjin Kim, **Donghyeon Han**, and Hoi-Jun Yoo, "C-DNN: A 24.5-to-85.8TOPS/W Complementary-Deep-Neural-Network Processor with Heterogeneous CNN/SNN Core Architecture and Forward-Gradient-based Sparsity Generation," *IEEE International Solid-State Circuits Conference*, Feb. 2023.
55. **Donghyeon Han**, Junha Ryu, Sangyeob Kim, Sangjin Kim, and Hoi-Jun Yoo, "MetaVRain: A 133mW Real-time Hyper-realistic-3D-NeRF Processor with 1D-2D Hybrid-Neural-Engines for Metaverse on Mobile Devices," *IEEE International Solid-State Circuits Conference*, Feb. 2023.

Index

© The Editor(s) (if applicable) and The Author(s), under exclusive license to Springer Nature Switzerland AG 2023
D. Han, H.-J. Yoo, *On-Chip Training NPU - Algorithm, Architecture and SoC Design*, https://doi.org/10.1007/978-3-031-34237-0

MIX
Papier aus verantwortungsvollen Quellen
Paper from responsible sources
FSC® C105338

If you have any concerns about our products,
you can contact us on
ProductSafety@springernature.com

In case Publisher is established outside the EU,
the EU authorized representative is:
Springer Nature Customer Service Center GmbH
Europaplatz 3, 69115 Heidelberg, Germany

Printed by Libri Plureos GmbH
in Hamburg, Germany